Challenges in Taste Chemistry and Biology

ACS SYMPOSIUM SERIES **867**

Challenges in Taste Chemistry and Biology

Thomas Hofmann, Editor
Universität Münster

Chi-Tang Ho, Editor
Rutgers, The State University of New Jersey

Wilhelm Pickenhagen, Editor
Dragoco Gerberding and Company

Sponsored by the
ACS Division of Agricultural and Food Chemistry

American Chemical Society, Washington, DC

Library of Congress Cataloging-in-Publication Data

Challenges in taste chemistry and biology / Thomas Hofmann, editor, Chi-Tang Ho, editor, Wilhelm Pickenhagen, editor ; sponsored by the ACS Division of Agricultural and Food Chemistry.

p. cm.—(ACS symposium series ; 867)

Includes bibliographical references and index.

ISBN 0-8412-3852-9

1. Flavor—Congresses. 2. Flavoring essences–Congresses. 3. Food—Odor—Congresses. 4. Taste—Congresses.

I. Hofmann, Thomas F., 1968- II. Ho, Chi-Tang, 1944- III. Pickenhagen, Wilhelm, 1939- IV. American Chemical Society. Meeting (224th : 2002 : Boston, Mass.) V. Series.

TP372.5.C43 2003
664′.07—dc22 2003057712

The paper used in this publication meets the minimum requirements of American National Standard for Information Sciences—Permanence of Paper for Printed Library Materials, ANSI Z39.48–1984.

Distributed by Oxford University Press

PRINTED IN THE UNITED STATES OF AMERICA

Foreword

The ACS Symposium Series was first published in 1974 to provide a mechanism for publishing symposia quickly in book form. The purpose of the series is to publish timely, comprehensive books developed from ACS sponsored symposia based on current scientific research. Occasionally, books are developed from symposia sponsored by other organizations when the topic is of keen interest to the chemistry audience.

Before agreeing to publish a book, the proposed table of contents is reviewed for appropriate and comprehensive coverage and for interest to the audience. Some papers may be excluded to better focus the book; others may be added to provide comprehensiveness. When appropriate, overview or introductory chapters are added. Drafts of chapters are peer-reviewed prior to final acceptance or rejection, and manuscripts are prepared in camera-ready format.

As a rule, only original research papers and original review papers are included in the volumes. Verbatim reproductions of previously published papers are not accepted.

ACS Books Department

Contents

Analytical Characterization and Structure–Activity Relationships

Taste-Active Peptides and Amino Acid Derivatives

Flavor Interactions

Indexes

Preface

Flavor is the simultaneous stimulation of the human senses (odor and taste) and is triggered by distinct molecules occurring in our daily foods. Although the consumer acceptance of foods is strongly influenced by both the aroma-active volatiles as well as the nonvolatile, taste-active compounds, flavor research in the past decades focused mainly on the aroma rather than on the taste compounds. Presently, the world of taste research seems to be split into two parts. One focuses on the chemical structures and sensory activities of food taste compounds. Recent advances in coupling high-performance liquid chromatography with human sensory analysis as well as with mass spectrometric and nuclear magnetic resonance spectroscopy have strongly improved our understanding of the importance of single tastants for overall food quality. Nevertheless, relatively little attention has been paid to the tastants that are not present in the foods per se, but that are generated during food processing from taste-less precursors. The other, more biochemically orientated research party, puts its research effort into the identification of taste receptors, signaling molecules, and transduction pathways on the protein as well as the gene level. Molecular–biological investigations as well as recent developments in sensory science have greatly improved our knowledge of physiology and the mechanism of human taste perception.

Unfortunately, the interchanges between these two disciplines of taste research are far away from being satisfactory. The two day symposium *Taste Research: Chemical and Physiological Aspects*, held at the 224th National American Chemical Society (ACS) meeting in Boston, Massachusetts, August 2002, was, therefore, intended to bring together the leading scientists of both parties from academia and industry and to provide a forum for discussion between researchers investigating the biochemistry, genetics, and physiology of human taste transduction–perception; chemists focusing on the identification and structure–activity relationships of the key compounds involved in the taste of our foods; and more applied industrial researchers who are interested in human sensory analysis, in taste–aroma interactions and in the possibilities of how to technologically optimize tastant formation during food processing. This book covers the breakthrough research and highlights of that symposium, which will give us a more comprehensive understanding of human taste perception. The chapters of this book will be of interest to food chemists, food processing companies, flavor houses, sensory analysts, and those focused on the physiological mechanisms of taste transduction and taste modification such as biochemists or pharmacists.

The editors acknowledge with great appreciation financial support from the ACS Division of Agricultural and Food Chemistry, Dragoco, International Flavors and Fragrances, Nestle, and Procter & Gamble. We also thank the ACS Books Department for encouraging this publication. Finally, we thank all the authors for their contributions and cooperation in the preparation of this book.

Thomas Hofmann
Institut für Lebensmittelchemie
Universität Münster
Corrensstrasse 45
D–48149 Münster
Germany
thomas.hofmann@uni-muenster.de

Chi-Tang Ho
Department of Food Science
Rutgers, The State University of New Jersey
New Brunswick, NJ 08901
ho@aesop.rutgers.edu

Wilhelm Pickenhagen
Dragoco Gerberding and Company AG
Corporate Research Division
Dragocostrasse
D–37603 Holzminden
Germany
Pickenhagen.Wilhelm@symrise.com

Challenges in Taste Chemistry and Biology

Chapter 1

Challenges in Taste Research: Present Knowledge and Future Implications

Thomas Hofmann[1], Chi-Tang Ho[2], and Wilhelm Pickenhagen[3]

[1]Institut für Lebensmittelchemie, Westfälische Wilhelms-Universität, Corrensstrasse 45, D–48149 Münster, Germany
[2]Department of Food Science, Rutgers, The State University of New Jersey, 65 Dudley Road, New Brunswick, NJ 08901–8520
[3] Dragoco Gerberding and Company, AG, D–37603 Holzminden, Germany

This chapter gives a brief survey on the chemistry of gustatory taste sensations, i.e. the perception of the basic tastes modalities sour, sweet, salty, bitter as well as umami, and lingual somatosensory sensitivity resulting from temperature and tactile stimulation as well as chemical activation of chemosensory receptors. Some of the present challenges facing researchers in taste chemistry and biochemistry are presented and future perspectives are discussed.

Italian wine... Swiss chocolate... French cuisine... much of our daily life quality is depending on the proper function of our sensory system. Although aided by smell and visual inspection, the final recognition and evaluation of food relies on chemoreceptive events in the oral space. This sensing system gives us the possibility to detect those foods containing essential nutrients such as salty tasting minerals, sweet tasting carbohydrates, and sweet or umami-tasting amino acids, but also to detect aversive bitter tasting compounds in order to warn about the potential ingestion of toxic or harmful chemicals such as, e.g. strychnine, amygdaline or nicotine. This functional duality of taste is reflected in a great diversity of different compounds that are capable of activating the gustatory system in the mouth.

The lingual flavor sensation during food delight is a complex convergence of at least two sensory modalities; (i) the gustatory taste sensation, i.e. the perception of the basic tastes modalities sour, sweet, salty, bitter as well as umami by activation of taste bud sensory cells, and (ii) lingual somatosensory sensitivity resulting from temperature and tactile stimulation as well as chemical activation of chemosensory receptors on the perigemmal fibers. An enormous number of investigations have been done in the past and in the present to understand the biochemistry of tastant recognition and signal transduction, as well as to structurally characterize the chemical species inducing the somato-gustatory responses perceived.

The basic taste modalities

Mammalian taste receptor cells are neuroepithelial cell clustered into sensory end organs called taste buds, that are distributed in distinct regions of the tongue and the palate epithelium. Each taste bud contains about 50-150 cells, including precursor cells, support cells, and taste receptor cells (*1*). On the apical surface taste receptor proteins, mounted on microvilli, provide the molecular specificity to the receptor cells which are innervated by afferent nerve endings that transmit information to the taste centers of the cortex through synapses in the brain stem and thalamus. Although these "chemosensors" in taste buds are monitoring the complex chemical environment in the oral cavity and generate signals that lead to diverse taste perceptions, individual taste stimuli, even within one taste tonality, often differ greatly in molecular size, lipophilicity, and pH optimum. Thus the taste system must utilize a diversity of mechanisms for ligand recognition and signal transduction to accommodate the structural diversity of the chemicals to be detected.

Taste transduction for complex organic molecules such as soluble carbohydrates, amino acids, and the heterogeneous collection of compounds that elicit perception of sweetness, bitterness, and umami taste is believed to include binding at specific G-protein-coupled receptors, GPCRs, on the taste cell plasma membrane (*2*). Modern molecular-biological technologies have enabled the identification of some key molecules in the plethora of GPCRs, ion channels, ligand-gated channels, enzymes and other signaling molecules triggering the downstream transduction events within taste cells.

Bitter taste

By scanning the genomic data bases near a promising bitter locus on mouse chromosome 6, a novel group of about 30 different GPCRs was discovered, the so-called T2R family (*3, 4*). After heterologous expression in a cell line, a mouse T2R (mT2R-5) was shown to respond to the bitter tasting cycloheximid, and a human (hT2R-4) and a mouse receptor (mT2R-8) responded to denatonium and 6-n-propyl-2-thiouracil (*5*). From *in situ* hybridization experiments it was concluded that a single taste receptor cell expresses a large number of T2Rs, suggesting that each cell may be capable of recognizing multiple tastants (*3*). But using functional imaging systems it could be shown that most bitter responsive taste cells, which presumably expressed T2Rs, were activated by only one out of five bitter compounds testes (*6*).

Furthermore, most T2Rs have been found to be co-expressed in the same subset of taste receptor cells of the tongue and palate ephithelium (*3*), suggesting that these taste cells are capable of responding to a broad array of bitter chemicals. T2Rs may also function as heterodimeric receptors to accommodate the great chemical multiplicity of bitter tastants in foods. This would explain how the great variety of organic molecules is interfering with the internal signaling system of the humans tongue. Bitter compounds in foods cover a wide range of chemical classes (Figure 1); for example, the structural diversity of bitter taste compounds in plant materials spans from flavanoids such as the bitter principle of grapefruit naringine (*7*), cyanogene glycosides such as (*R*)-amygdaline in bitter almonds (*8*), bitter alkaloids such as caffeine in coffee, tea, and guarana (*7*), bitter tasting amino acids such as L-tryptophane (*7*), or lipid diols such as falcarindiol, the compound imparting bitterness to stored carrots (*9*). Other bitter compounds are not present in the raw food *per se*, but are formed from taste-less precursors during food processing such as enzymatic reactions; for example, the bitter peptide QNKIHPFAQTQSLVYPFPGPIP is formed from casein proteolyses during ripening of cheddar cheese (*10*), the bitter tasting 9,10,13-trihydroxy-(*E*)-11-octadecenoic acid is formed from linoleic acid peroxidation in oats (*11*), the bitter divinylglycol is formed during off-fermentation of wines (*7*). Additional bitter compounds are formed during thermal food processing from precursors of diverse chemical classes; for example, iso-α-humolone is generated during beer brewing from its less bitter, hop-derived terpenoid precursor α-humulone (*12*), the bitter goitrine is formed during cooking of Brussels sprouts from a glucosinolate (*13*), bitter tasting diketopiperazines such as cyclo(Phe-Val) are produced during roasting of cacao beans by dimerization of amino acids, peptides, and proteins (*14*), and the cyclopenta[*b*]azepinone CPA is generated in roasted malt by Maillard reactions from hexoses and the secondary amino acid L-proline (*15*).

Besides the GPCRs, alternative ways in detecting some bitter compounds are reported; for example, some bitter peptides with amphophilic character interact directly with G-proteins, probably by virtue of a structural similarity to the G-protein binding site of the receptor (*16*). Bypassing the receptor, quinine is reported to directly activate G-proteins (*17*) and caffeine was found to block a phosphodiesterase, thus inducing activation of a guanylate cyclase resulting in a transient increase of cGMP (*18*).

Sweet taste

For much of the world, honey appears to have been the main source of sweetness until about six centuries ago, when sucrose became widely available in European countries. Commercial supplies of sucrose were derived from the sap of sugar cane, although this carbohydrate is now also economically

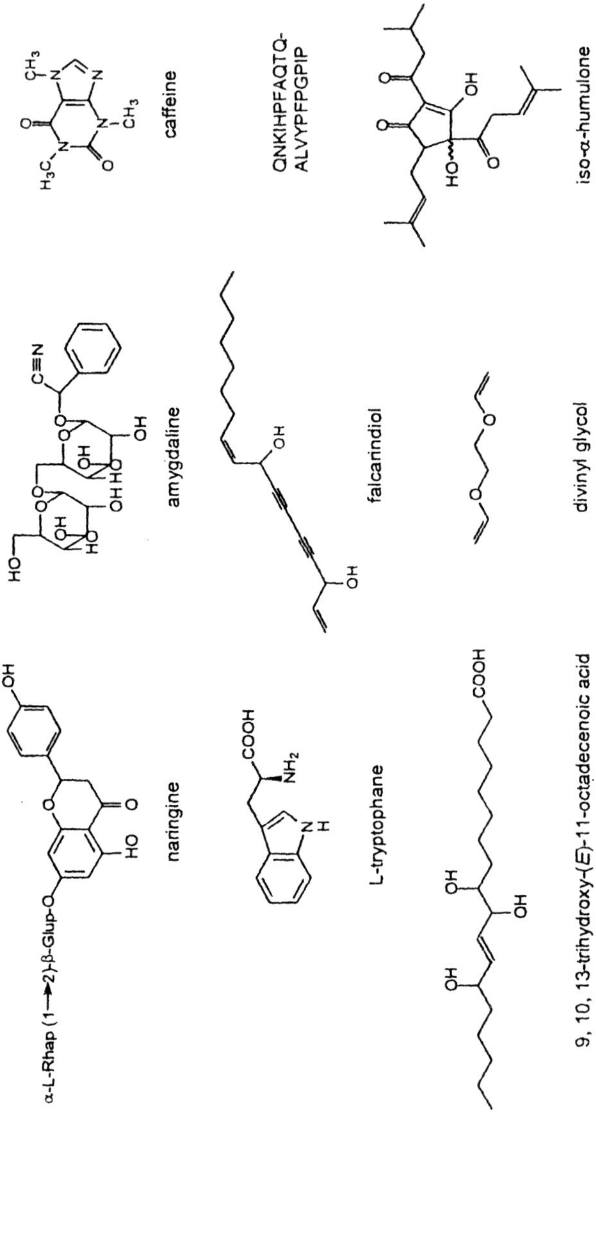
caffeine
QNKIHPFAQTQ-
ALVYPFPGPIP
iso-α-humulone
amygdaline
falcarindiol
divinyl glycol
α-L-Rhap (1⟶2)-β-Glup-O
naringine
L-tryptophane
9, 10, 13-trihydroxy-(E)-11-octadecenoic acid

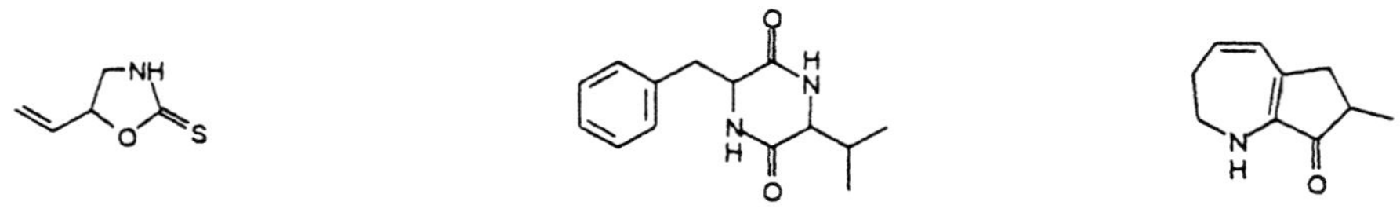

goitrine cyclo(Phe-Val) CPA

Figure 1. Structural diversity of bitter tasting molecules identified in raw and processed food.

produced from sugar beet and, to a lesser extend, from sugar maple. Sweet taste responds to soluble carbohydrates, such as sucrose, guiding high-calorie intake, but also to a variety of other non-nutritive chemicals such as, e.g. glycyrrhizine, D-tryptophane, neohesperidine dihydrochalcone, and the artificial sweeteners aspartame and saccharin (Figure 2), inducing a strong pleasant hedonic effect (*19*). In the last decade, considerable efforts have been made by chemists to deduce from a library of sweet-tasting compounds common structural features and receptor binding-site models, which could be successfully used to design new high-potency sweeteners such as, e.g. lugduname which is 230000 times more sweet than sucrose (*20*).

Aimed at defining sweet receptors more directly on a molecular-biological level, the receptor T1R3, possessing a large N-terminal domain like the orphan receptors T1R1 and T1R2, was recently found in taste buds of the anterior, lateral and posterior tongue (e.g. *21, 22*). As the gene *Tas1r3* is the only GPCR-coding gene at the *Sac* locus, a sweet taste-related location on sweet-receptor genes, its product T1R3 was suggested as a candidate for a sweet receptor. The T1R3 is expressed in many taste cells which also express T1R2 suggesting that the two receptors may serve as a common function, probably in form of heterodimers. This could very recently be confirmed as the T1R3 receptor was shown not to respond to sweeteners on its own, but responses were obtained only after co-expressing T1R3 together with T1R2 (*23, 24*). This example showed the first functional sweet receptor found in mammals as a heterodimer.

Umami taste

The taste modality "umami", a term derived from the Japanese "umai" ("delicious"), designates a pleasant taste sensation imparted by L-glutamate and, even more pronounced, by mono sodium glutamate (MSG, Figure 3) in savory foods such as, e.g. meet extracts, soy sauce, protein hydrolysates and ageing cheese. Until recent molecular-biological findings, only sweet, sour, salty, and bitter have been regarded as the basic taste qualities, and the existence of umami taste as an additional primary taste has been discussed controversially in the literature. In recent molecular-biological investigations a taste-specific truncated form of the metabotropic glutamate receptor mGluR4 present in the brain was identified from rat taste buds and was suggested as a taste receptor for L-glutamate (*25*). The identification of this novel G-protein coupled receptor in taste tissue provides evidence for the existence of glutamate-like taste as a basic taste quality. Several compounds in addition to glutamate elicit an umami taste including inosine-5'-monophosphate (5'-IMP) and guanosine-5'-monophosphate (5'-GMP, Figure 3) (*26*). These purine-5'-ribonucleotides occur in many savory foods such as meat, fish, seafood, and mushrooms, and enhance the flavor and mouthfeel including the impression of creaminess and viscosity of savory dishes. Therefore, they are widely used as ingredients and taste-enhancers in culinary products and snacks, e.g. soups, sauces and savory seasonings.

Recently, *N*-lactoyl-L-glutamate (Figure 3), the condensation product of lactic and glutamic acid, was reported to possess glutamate-like, MSG-type organoleptic properties (*27*). It was shown that the glutamate derivative evokes a bouillon-like umami taste, gives a long-lasting and mouth-watering sensation and has flavour-enhancing characteristics, although weaker compared to MSG. In agreement with the sensory qualities of *N*-lactoyl glutamate assessed by the human taste panel, a comprehensive molecular-biological investigation demonstrated that *N*-Lactoyl-L-glutamate activated the rat mGlu4a receptor and inhibited the binding of L-AP4, a selective L-glutamate agonist for mGlu4 receptors (*28*).

But is taste mGluR4 the only umami receptor? The taste mGlu4 shows structural similarity and approximately 27% sequence identity with the taste receptor proteins T1R1 and T1R2 (*29*). T1Rs are differentially expressed in the tongue, with T1R1 expressed predominantly in taste buds of fungiform papillae and palate, and T1R2 expressed almost exclusively in circumvallate and foliate taste buds. In contrast to the bitter taste mediated by an unrelated family of the T2Rs (*3, 4*), and sweet taste mediated by a T1R2/T1R3 heterodimer to function as a sweet receptor (*23, 24*), T1R1 and T1R3 were recently found to form a universal heterodimeric L-amino acid (umami) sensor (*30, 24*). In summary, the T1R gene family encodes three conserved GPCRs that function as heterodimers in heterologous expression systems and form either a sugar receptor in the T1R2/T1R3 combination, or a general L-amino acid receptor in the T1R1/T1R3 combination.

In addition to their characteristic sensory quality, another peculiar effect of umami tasting compounds is their mutual taste synergism. The synergistic effects between L-glutamate and purine-5'-ribonucleotides and analogues of both groups have been observed more than 35 year ago (*31, 32*). Systematic investigations discovered an exponential increase in the glutamate-like taste intensity of a aqueous solution of MSG when 5'-IMP was added. According to the results of these studies the taste activity of 5'-GMP is 2.3 fold the effect of 5'-IMP, while monosodium aspartate possesses about 7% of the efficacy of monosodium glutamate. Molecular-biological investigations recently succeeded in confirming the synergistic effect of purine-5'-ribonucleotides on the umami taste of MSG at the taste receptor level. Human T1R1/T1R3 heterodimeric receptors, made up by coexpression of the C-family G protein-coupled receptor T1R1 and the related taste specific receptor T1R3, were demonstrated to respond to the umami-type taste stimulus L-glutamate (*24*). Although 5'-IMP and 5'-GMP, two hallmarks of taste enhancement, alone did not activate human T1R1/T1R3, these purine-5'-ribonucleotides strongly potentiated the L-glutamate induced T1R1/T1R3 receptor response. The key role of the T1R1/T1R3 dimer in responding to umami-like tastants was further confirmed

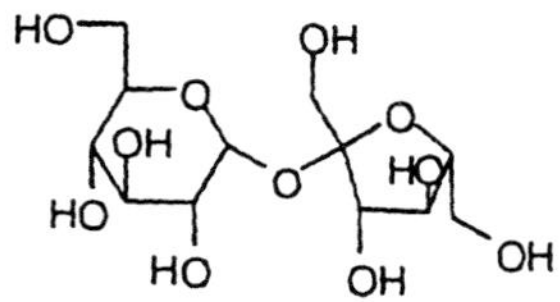

sucrose

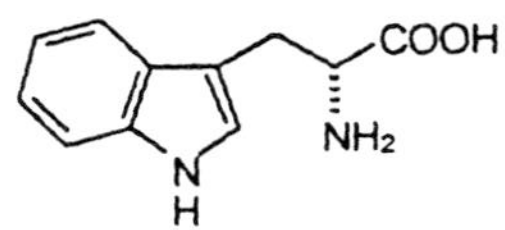

D-tryptophane

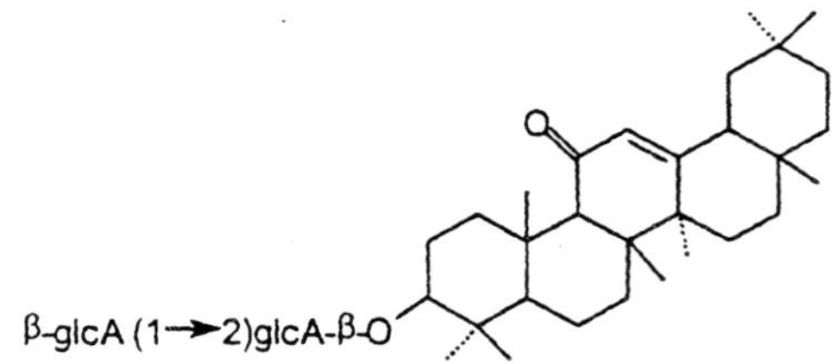

glycyrrhizine

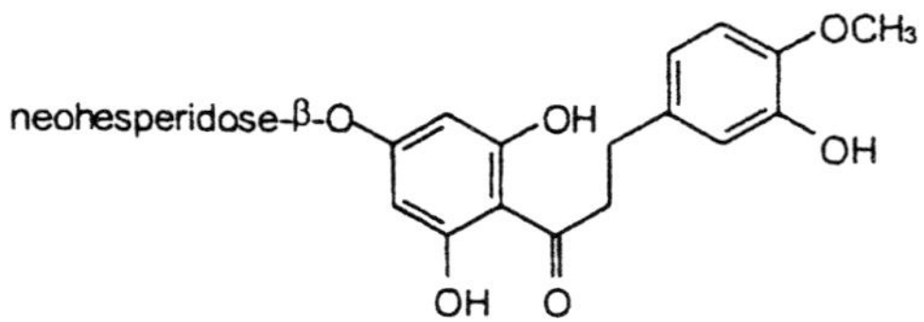

neohesperidine dihydrochalcone

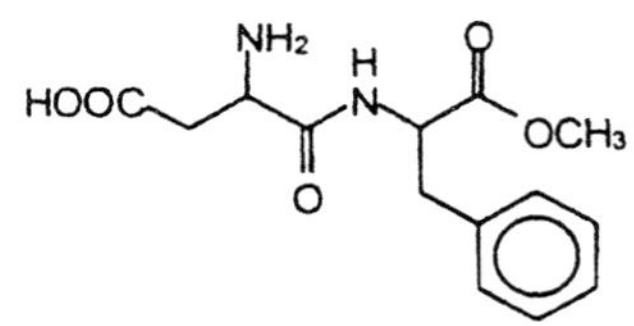

aspartame

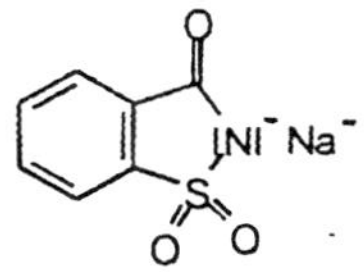

saccharin

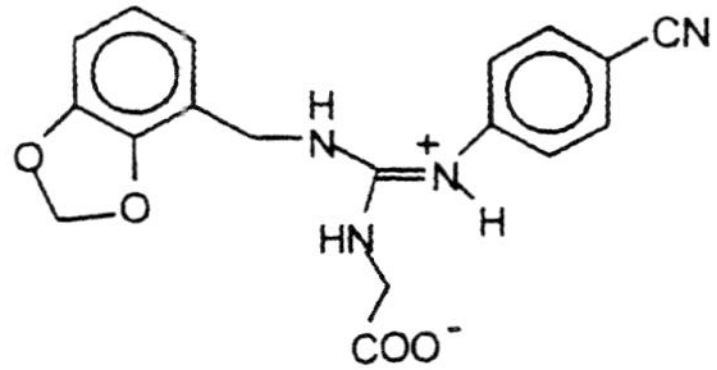

lugduname

Figure 2. Chemical structures of sweet tasting molecules in foods and sweeteners used in food applications.

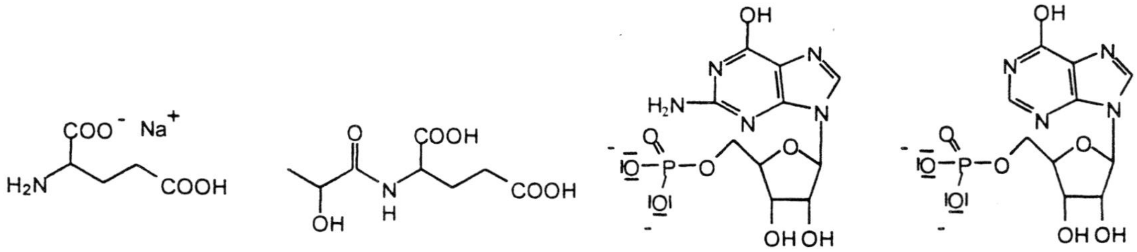

MSG N-lactoyl glutamate 5'-GMP 5'-IMP

Figure 3. Chemical structures of compounds exhibiting umami taste.

as cytidin-5'-monophosphate, which does not enhance human umami-type taste perception, showed no effect (*24*).

Signal transduction for bitter, sweet and umami taste

A plethora of GPCRs, ion channels, ligand-gated channels etc. serve as receptors for taste compounds and trigger the downstream transduction cascade within taste receptor cells. Aimed at identifying key molecules in the signal transduction of bitter taste, members of the T2R family were found to be co-expressed with the α-subunit of the G-protein gustducin, a taste-specific signaling protein inducing the decrease of the cellular concentration of cyclic nucleotides by activating a taste-specific phosphodiesterase (*33*). Consistent with the proposed role of α-gustducin in bitter transduction, the phosphodiesterase (PDE) from bovine taste tissue was shown to be activated *in vitro* by gustucin as well as by transducin, the homologous G protein of the visual system also present in taste buds (*34*). Knockout mice lacking α-gustducin showed significantly decreased sensitivity for the bitter compounds denatonium and quinine, and also for the sweet tasting sucrose and saccharin (*35*). In consequence, gustducin may be involved in both bitter and sweet response (*36*). In contrast to the robust activation of gustducin by bitter stimulated taste receptors, neither sucrose, nor artificial sweeteners activate gustducin in the presence of taste receptor-containing membranes. However, a number of artificial sweeteners competitively inhibited bitter-receptor activation of gustducin suggesting that these compounds may act as bitter antagonists (*37*).

In parallel to the α-gustducin/PDE pathway, an alternative transduction cascade is stimulated simultaneously by the GPCR activation. The gustducin subunits Gβ3 and Gγ13, which have been found colocalized with α-gustducin to give a heterotrimeric gustducin (38), are able to activate phospholipase Cβ2 which in turn induces the liberation of inositol-1,4,5-triphosphate (IP_3). After activation of IP_3-receptors of intracellular Ca^{2+} stores, the concentration of Ca^{2+} ions increases. But the channels responsible for a change in membrane potential, resulting in a synaptic activation and nerve exitation, need to be identified. These data suggest that sweet and bitter stimuli activate taste cells through at least two transduction pathways, of which one involved an increase in cyclic nucleotides, the other an increase of IP_3 (*39*). It has also been suggested that the recognition of sugars and artificial sweeteners is mediated by different signaling molecules and transduction pathways (*40*).

These informations would imply that a single taste receptor cell (TRC) could be mounted with various taste receptors and could be therefore broadly tuned to recognize a multitude of different ligands. This implies that an individual cell might express many different taste receptors that couple and transduce the signal from a given receptor type (*40*). In contrast to this hypothesis of broadly tuned TRCs, very recent investigations however argued for a conserved signaling cascade that is shared by all three receptor types for recognizing sweet, umami and bitter compounds (*41*). The TRC specific ion channel TRPM5, recently suggested to play a role in bitter taste transduction (*42*), could be shown not only to be expressed in bitter-responsive T2R-containing cells, but also in T1R positive cells. Carefully planned knockout-mice studies established that TRPM5 and the phospholipase C (PLCß2), a major signaling effector of GPCRs, are co-expressed with T1Rs and T2Rs and are vital for sweet, umami, and bitter taste transduction (*41*), whereas salty and sweet response was not affected. Based on these studies, it was proposed that sweet, amino acid (umami), and bitter taste converge on a common transduction channel via PLC. Tastant activation of T1R or T2R receptors would stimulate G proteins, and in turn PLCß2. The stimulation of PLC, either directly, or indirectly would lead to the gating of TRPM5 and the generation of a depolarizing receptor potential (*41*).

In addition, mice engineered to rescue PLCß2 function exclusively in bitter-receptor expressing cells responded to bitter tastants but do not taste sweet or amino acid stimuli. That means that bitter is encoded independently of sweet and amino acids. Thus taste coding seems to be strictly separated at the cellular level, i.e., taste receptor cells are not broadly tuned across different taste modalities, but mediate only one of these three qualities (*41*).

Sour and salty taste

Two taste qualities detect ions in the oral space; salty and sour. Taste transduction of simple inorganic salts and acids involves altered permeation of the receptor cell membrane by direct interaction of ions such as Na^+, K^+, or H^+ with particular ion channels (*39*). The resulting receptor currents in taste bud cells stimulate neurotransmitter release to excite sensory afferents, ultimately leading to perceptions such as "salty" or "sour".

Though several mechanisms have been proposed for the transduction of Na^+ salts, the only molecular components that have been identified unequivocally as contributing to this transduction mechanism are the family of ephitelial-type sodium channels (ENaCs) (*1, 39, 36*).

Acids have been shown in TRCs to block ion channels, permeate ion channels, activate ion channels, alter transporter function and change intracellular pH (*36*). Despite of the variety of effects and potential targets, there has been only little success until recently at characterizing the molecular species involved in the transduction machinery.

Somatosensory stimuli during food ingestion

Besides the perception of the basic tastes modalities sour, sweet, salty, bitter as well as umami by activation of taste bud sensory cells, lingual somatosensory sensitivity resulting from temperature and tactile stimulation as well as chemical activation of chemosensory receptors on the perigemmal fibers contribute to the somato-gustatory sensation during food intake.

Pungency

Trigeminal nerve endings in the mouth contribute to flavor through the sensory modalities of touch, thermal sensation and pain as mediated by some spices in Western cuisine, for example, chilli peppers, pepper and ginger with their active key ingredients capsaicin, piperin and gingerol, respectively (Figure 4).

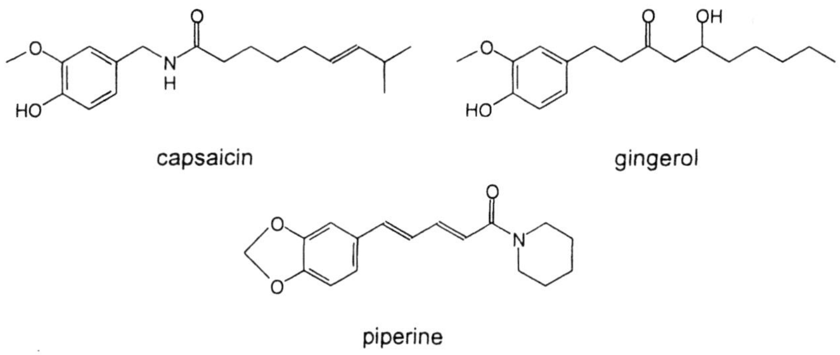

Figure 4. Chemical structures of pungent key compounds in chilli peppers, ginger, and pepper.

The best-characterized example of chemically induced trigeminal sensation is the pungency produced by "hot" chilli peppers. Capsaicin, the most pungent ingredient in chilli peppers, elicits a sensation of burning pain by selectively activating peripheral terminals of a subgroup of sensory neurons, called nociceptors, that convey information about noxious stimuli to the central nervous system. A more detailed understanding of the molecular nature of capsaicin action was recently obtained through gene cloning and characterization of two ion channels that function as the principal heat sensors on mammalian somatosensory neurons. The vanilloid receptor channel subtype

1 (VR1) was found to be activated by moderate temperatures above 43 °C as well as by vanilloid compounds including the pungent capsaicin (*43*), whereas the vanilloid-receptor-like channel type 1 (VRL-1) was shown to respond to higher temperatures above 50 °C (*44*).

Cooling sensation

Some compounds commonly found in spices, food and beverages also elicit sensations other than pain; for example, menthol and the related compounds (-)-menthyl lactate and N-ethyl-p-menthane-3-carboxamide (Figure 5) stimulate a subclass of thermal nerve endings to produce cooling.

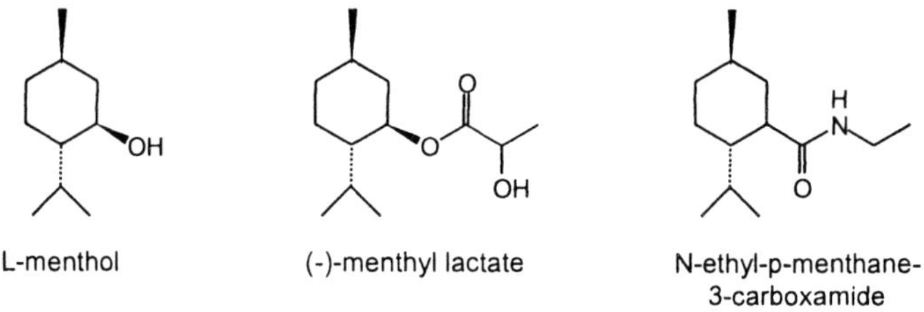

Figure 5. Chemical structures of compounds with physiological cooling activity.

Recent investigations succeeded to identify a cDNA encoding a receptor that responds to the cold "mimic" menthol (*45*). In addition, this menthol-activated ion channel, which the authors named cold–menthol receptor type 1, CMR1, was also activated by cold temperatures in the range of 8–30 °C. For the first time, these data provide a molecular explanation that pungent compounds and heat as well as cooling compounds and cold are detected by the same molecular entity on primary afferent neurons of the somatosensory system. Beginning with a bioinformatics approach new members of the TRP family were isolated, including TRPM8 (*46*). *In situ* hybridization experiments and functional studies then showed that this channel, which seems to be identical to CMR1, is a cold receptor expressed in a subset of somatosensory neurons that detect temperature and noxious stimuli.

Tingling sensation

other compounds such as some unsaturated alkylamides (Figure 6) found in the fruit of Xanthoxylum, Szechuan pepper, and other plants, activate cold-sensitive and tactile nerve endings inducing an intense tingling sensation (47).

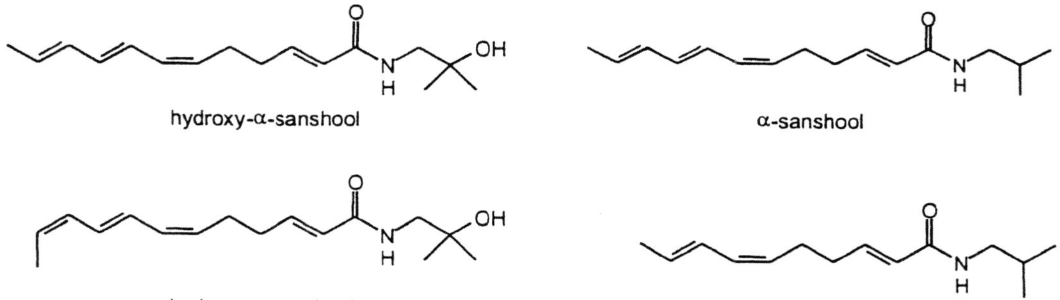

hydroxy-α-sanshool

α-sanshool

hydroxy-ε-sanshool

spilanthol

Figure 6. Structures of naturally occurring unsaturated alkylamides (47)

Sensory evaluation of 25-50 μg of hydroxy-α-sanshool, applied directly on the tongue did not elicit the same warming sensations as do capsaicins. After a delayed onset of 30-90 sec, the tingling sensation is more similar to a mild electric shock (5-7 V) or a carbonated solution applied to the tongue (*47*). At higher concentrations, the sensation is painful. In addition, the sensation of evaporative cooling of the tongue during inhalation is enhanced. Extracellular nerve recordings obtained from the lingual nerve of rats revealed that the primary pungent compound, hydroxy-α-sanshool, altered the levels of spontaneous activity in cool-sensitive fibers as well as inducing activity in tactile fibers, cold nociceptors, and silent fibers that were insensitive to innocuous thermal or tactile stimuli. Moreover, tactile or thermal sensitivity was induced in fibers that were initially insensitive to touch or cooling. The neuronal distribution of sensitivities to capsaicin and to hydroxy-α-sanshool indicate that these alkylamides affect neurons mediating innocuous sensations (*48*).

The catalog of chemical trigeminal stimuli is further extended by carbon dioxide (CO_2) inducing an oral irritant sensation. The ingestion of carbonated beverages such as, e.g. soda, beer, and champagne, induces a pleasurable and sought-after tingling sensation as if the bubbles of carbon dioxide are popping inside the mouth and is believed to be one of the major hedonic components contributing to the consumers attraction of carbonated drinks (*49*). It has been debated whether the oral sensation produced by carbonated beverages is preliminary chemogenic or rather mechanical in nature because of bursting CO_2 bubbles (*50*). Several studies argued against the mechanical hypothesis. Tingling, mouth-burn, pricking elicited by carbonated water under normal atmospheric conditions were essentially unchanged when subjects ingested the carbonated water under hyperbaric conditions in which bubble formation as prevented (*51*). Furthermore, subjects consistently reported a "burning and tingling numbness" aftersensation long after the carbonated water had been expectorated (*50*). Since mountaineers, after taking the carbonic anhydrase inhibitor acetazolamide for mountain sickness, reported that carbonated beverages lacked the "fizziness" and beer tasted like "dishwater" (*52*), the carbonic anhydrase-catalyzed conversion of CO_2 to carbonic acid was suggested as a key mechanism for the tingle of carbonated drinks:

$$CO_2 + H_2O \rightleftharpoons H_2CO_3 \rightleftharpoons HCO_3^- + H_2O$$

After conversion of CO_2, the carbonic acid induce the chemical exitation of intraoral nociceptors. The nociceptors, in turn, excite trigeminal neurons involved in signaling oral irritation to the lingual epithelium (*53, 54*). In recent psychophysiological experiments, it could be demonstrated that topical application of the carbonic anhydrase inhibitor acetazolamide to the tongue reduced the perceived intensity of the sensation induced by carbonated water in a concentration-dependend manner (*54*). This confirmed the idea that the sensation elicited by carbonated water on the tongue is caused by the exitation of lingual nociceptors via carbonic anhydrase catalyzed conversion of CO_2 to carbonic acid. This mechanism of signaling oral irritant sensation by carbonated beverages was further supported by recent investigations (*53*).

Salivary carbonic anhydrase isoenzyme VI (CA VI) is secreted from the large salivary glands and also from von Ebner's glands, small serous glands lying among lingual muscle fibers in the posterior tongue, to the buttom of the treches surrounding the circumvallate and foliate papillae and was immunochemically identified in taste buds and in taste pores. The CA VI appears to contribute to taste functionality by protecting taste receptor cells from apoptosis (*55*). Being well in line with these data, it could be recently shown that patients with taste loss have a lower salivary CA VI concentration and more apoptosis in taste receptor cells (*56*).

Astringency

Another oral sensation, the astringent sensation, may be sensorially described as a long-lasting puckering, shrinking, rough or and drying sensation on the tongue, teeth and oral mucosa, which is chemically induced by polyphenols (Figure 7) and aluminium salts.

The tactile sensation of astringency is thought to be perceived by touch via mechanoreceptors (*17*). For water-soluble phenols, molecular weight between 500 and 3000 Da were reported to be required (*57*). These polyphenols are discussed to decrease saliva lubrication and have a strong affinity for precipitating proline-rich salivary proteins or to form unprecipitated complexes (*58*). As the consequence, the effectiveness of salivary lubrication is decreased, and astringency is perceived as the friction between two "non-lubricated" surfaces. It has been speculated that astringency is the friction perceived when oral lubrication is reduced upon binding of astringent compounds with salivary proteins (*59, 60, 61*). It could be shown that subjects with low saliva flow rate evaluated astringency of red wine (*62*) and black tea (*63*) significantly higher than high flow subjects over eight successive sips. In conflict with these data, no effect of salivary flow on astringency perception was observed for astringent reference compounds such as, e.g. mono-, di- and trimeric flavanols (*64*) and hydroxy benzoic acids (*65*), thus indicating that the key compounds for this sensation in wine and tea are not yet known. Also interactions with oral surfaces have not yet been studied and may play a role in the astringent sensation.

Chemoreception of fatty acids

In contrast to those chemical stimuli inducing sour, salty, sweet, bitter, umami, pungent, tingling and cooling sensations, , the only sensory cue for dietary fat was until recently assumed to be its texture and "mouth-feel". However recent evidence suggests that free fatty acids contained in fats and oils, or liberated from the triglycerides during fat ingestion, may affect taste receptor cells (*66, 67*). Lingual lipase (*68*), fatty acid transporters (*69*), and lipocalin-

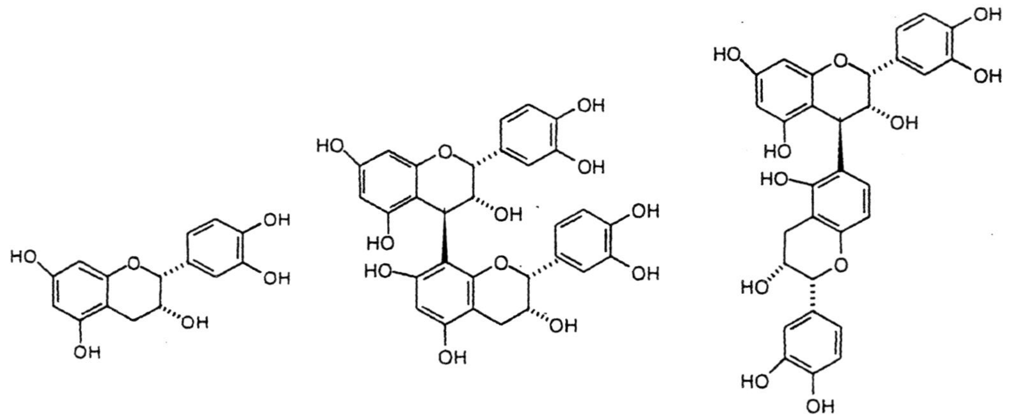

(-)-epicatechin procyanidine B-2 procyanidine B-5

Figure 7. Chemical structures of (-)-epicatechin, procyanidine B-2 and procyanidine B-5 eliciting an astringent sensation.

related von Ebner's gland proteins (*70*) have been identified in the oral cavity and are proposed to facilitate the release and transport of free fatty acids into? the area of the taste buds. Recently, strong evidence was found that (*Z*)-configurated, polyunsaturated fatty acids as primary signaling molecules in fat activate taste cells via an inhibition of Shaker Kv1.5-like delayed rectifying K (DRK) channels. More than 95% of taste cells responded to linoleic acid, linolenic acid, and arachidonic acid, but neither to myristic acid, nor to oleic acid (*71*). Further structure/activity investigations revealed that the corresponding methylesters were not effective at all, and conjugated linoleic acid with (*E,Z*)- or (*Z,E*)-configuration were only moderately effective demonstrating that a minimum of one [*Z*]- configurated double bond is required for the activity.

As putative mechanisms in the gustatory response to fatty acids, it is discussed that fatty acids may directly inhibit DRK channels to depolarize taste receptor cells leading to neurotransmitter release onto afferent nerve fibers, or after being transported through the lingual epithelium may directly affect basolateral DRK channels or trigeminal nerve fibers (*66*). This activation may represent a gustatory cue for fat as well as have profound implications for the ability of fats to modify the taste response of other sapid molecules. Animal trials showed that the peripheral responsiveness to fatty acids is correlated with the dietary fat preference. It was suggested that the fatty acid mediated inhibition of DRK channels may be a universal mechanism for the chemoreception of fat. Interestingly fatty acid mediated taste cell activation via interaction with DRK channels seems to be limited to the essential fatty acids.

Experiments demonstrating that oral fat exposure could lead to postprandial changes in lipid metabolism are supportive of there being a "taste" of fat (*72*). Numerous physiological studies have demonstrated that fats alter the hedonic response to other taste stimuli, in particular, sweet tastants (*73*). Therefore, there may be two effects of fatty acid inhibition of DRK channels in taste cells, it may serve as the transduction mechanism for dietary fats, and it may effect or modulate the gustatory responsiveness to other taste stimuli. Interestingly, individuals with the ability to taste the bitter substance 6-n-propylthiouracil (PROP), the so-called "super-tasters" (*74*), have the ability to discriminate the fat content of salad dressings (10% versus 40%) whereas PROP-insensitive "non-tasters" do not (*75*). Additional experiments will be, however, required to determine the precise role this transduction pathway plays in taste.

Future implications

Much effort has been made towards the characterization of receptor proteins and other signaling molecules of transduction within taste receptor cells. The sequencing of the receptor genetic code opens the way to study the corresponding proteins, to map their structure, and to achieve a better understanding of how the molecular machinery within taste GPCRs is working.

Although extensive investigations have been focused on the receptor proteins and signaling elements as well as on the molecular structure of the chemical stimuli inducing these taste responses, the pioneering spirit of taste researchers is still needed to tackle the many open open questions which have to be answered in future investigations: (i) what are the structures and sensory activities of trace taste compounds in foods, in particular those, which are not present in foods *per se*, but are formed during food processing from taste-less precursors, for example during roasting of coffee and cacao beans, fermentation of wine grapes, or cooking of beef broth etc., (ii) what are the structures of the missing key molecules involved in molecular recognition and signal transduction in taste cells, and (iii) how its it possible to stabilize receptor proteins without loosing functionality outside living cells. The answers on these questions will open the possibility to objectify the taste quality of foods on a molecular level, to characterize binding-site structures of receptors, to develop advanced techniques of drug design, and will be the basis for the development of long-term stable "bioelectronic tongues" using tastant-specific recombinant receptor proteins which are arranged in natural or synthetic bilayers and are contacted or interfaced to appropriate transducers.

From an industrial point of view, this will be the scientific basis for the construction of chemicals that activate or inhibit a receptor protein, thereby enhancing or inhibiting specific taste modalities. Thus it might become possible to develop organic molecules that enhance sweetness. This might be helpful for the development of low-calorie, sugar-reduced foods without the need of artificial sweeteners, which can show bitter and metallic off-taste at higher concentrations. The discovery of the salt and umami enhancing compounds will be a big step towards manufacturing of low-sodium foods for patients with hypertension, and might open new avenues for the production of umami-type tasting, low-glutamate savory foods for persons suffering from Chinese restaurant syndrome. Thus these developments in taste research will have implications in diverse fields, from food processing to clinical medicine.

References

(1) Lindemann, B. *Physiol. Rev.* **1996**, *76*, 719-766.

(2) Herness, M.S.; Gilbertson, T.A. *Annu. Rev. Physiol.* **1999**, *61*, 873-900.

(3) Adler, E.; Hoon, M.A.; Mueller, K.L.; Chandrashekar, J.; Ryba, N.J.P.; Zuker, Ch.S. *Cell* **2000**, *100*, 693-702.

(4) Matsunami, H.; Montmayeur, J.-P.; Buck, L. *Nature* **2000**, *404*, 601-604.

(5) Chandrashekar, J.; Mueller, K.L.; Hoon, M.A.; Adler, E.; Feng, L.; Guo, W.; Zuker, Ch.S.; Ryba, N.J.P. *Cell* **2000**, *100*, 703-711.

(6) Caicedo, A.; Roper, S.D. *Science* **2001**, *291*, 1557-1560.

(7) Belitz, H.D.; Wieser, H. *Food Rev. Int.* **1985**, *1*, 271-354.

(8) Hanssen, E.; Sturm, W. *Z. Lebensm. Unters. Forsch.* **1967**, *69*, 134.

(9) Czepa , A.; Hofmann, T. *J. Agric Food Chem.* **2003**, in press.

(10) Richardson, B.C.; Creamer, L.K. *NZJ Dairy Sci. Technol.* **1973**, *8*, 46.

(11) Biermann, U.; Wittmann, A.; Grosch, W. *Fette, Seifen, Anstrichmittel*, **1980**, *82*(6), 236-240.

(12) Palamand, S.R.; Aldenhoff, J.M. *J. Agric. Food Chem.* **1973**, *21*, 534.

(13) Fenwick, G.R.; Griffith, N.M. *Z. Lebensm. Unters. Forsch.* **1981**, *172*, 90-92.

(14) Pickenhagen, W.; Dietrich, P.; Keil, B.; Polonwsky, J.; Nonaille, F.; Lederer, E. *Helv. Chim. Acta*, **1975**, *58*, 1078-1086.

(15) Tressl, R.; Helak, B.; Köppler, H.; Rewicki, D. *J. Agric. Food Chem.* **1985**, *33*, 1132-1137.

(16) Spielman, A.L.; Huque, T.; Whitney, G.; Brand, J.G. In: *Sensory Transduction* (D.P. Corey, S.D. Roper, Eds.), The Rockefeller University Press, New York,1992, 307-324.

(17) Naim, M.; Seifert, R.; Nürnberg, B, Grünbaum, L, Schultz, G. *Biochem. J.* **1994**, *297*, 451-454.

(18) Rosenzweig, S.; Yan, W.; Dasso, M.; Spielman, A.I. *J. Neurophysiol.* **1999**, *81*, 1661-1665.

(19) Ganchrow, J.R.; Steiner, J.E.; Daher, M. *Infant Behav. Dev.* **1983**, *6*, 189-200.

(20) Nofre, C.; Tinti, J.M. *Food Chem.* **1996**, *56*, 263-274.

(21) Max, M.; Shanker, Y.G.; Huang, L.; Rong, M.; Liu, Z.; Champagne, F.; Weinstein, H.; Damak, S.; Margolskee, R.F. *Nat. Genet.* **2001**, *28*, 58-63.

(22) Montmayeur, J.P.; Liberles, S.D.; Matsunami, H.; Buck, L.B.A. *Nature Neurosci.* **2001**, *4*, 492-498.

(23) Nelson, G.; Hoon, M.A.; Chandrashekar, J.; Zhang, Y.; Ryba, N.J.P.; Zuker, C.S. *Cell* **2001**, *106*, 381-390.

(24) Li, X.; Staszewski, L.; Xu, H.; Durick, K.; Zoller, M.; Adler, E. *Proc. Natl. Acad. Sci.* USA, **2002**, *99*, 4692-4696.

(25) Chaudari, N.; Landin, A.M.; Roper, S. D. *Nat. Neurosci.* **2000**, *3*, 113-119.

(26) Kuninaka, A. In: *Symposium on Foods: The Chemistry and Physiology of Flavors* (Schultz H. W., Day E. A., Libbey L. M., eds.). AVI Publishing Company, Westport, Connecticut, 1967, pp. 515-535.

(27) Frérot, E.; Escher S., D. *WO 97/04667* (in French), **1997**.

(28) Monastyrskaia, K.; Lundstrom, K.; Plahl, D.; Acuna, G.; Schweitzer, C.; Malherbe, P.; Mutel, V. *Brit. J. Pharmacol.* **1999**, *128*, 1027-1034.

(29) Hoon, M.A.; Adler, E.; Lindemeier, J.; Battey, J.F.; Ryba, N.J.P.; Zuker, C.S. *Cell* **1999**, *96*, 541-551.

(30) Nelson, G.; Chandrashekar, J.; Hoon, M.A.; Feng, L.; Zhao, G.; Ryba, N.J.P.; Zuker, C.S. *Nature* **2002**, *416*, 199-202.

(31) Yamaguchi, S. *J. Food Sci.* **1967**, *32*, 473-478.

(32) Yamaguchi, S.; Yoshikawa, T.; Ikeda, S.; Ninomiya, T. *J. Food Sci.* **1971**, *36*, 1761-1765.

(33) McLaughlin, S.K.; McKinnon, P.J.; Margolskee, R.F. *Nature* **1992**, *357*, 563-569.

(34) Ruiz-Avila, L.; McLaughlin, S.K.; Wildman, D.; McKinnon, P.J.; Robichon, A.; Spickovsky, N.; Margolskee, R.F. *Nature* **1995**, *376*, 80-85.

(35) Wong, G.T.; Gannon, K.S.; Margolskee, R.F. *Nature* **1996**, *381*, 796-800.

(36) Gilbertson, T.A.; Damak, S.; Margolskee, R.F. *Curr. Opt. Neurobiol.* **2000**, *10*, 519-527.

(37) Ming, D.; Ninomiya, Y.; Margolskee, R.F. *Proc. Natl. Acad. Sci.* USA **1999**, *96*, 9903-9908.

(38) Huang, L.; Shanker, Y.G.; Dubauskaite, J.; Zheng, J.Z.; Yan, W.; Rosenzweig, S.; Spielman, Al.; Max, M.; Margolskee, R.F. *Nat. Neurosci.* **1999**, *2*, 1055-1062.

(39) Lindemann, B. *Nature* **2001**, *413*, 219-225.

(40) Margolskee, R.F. *J. Biol. Chem.* **2002**, *277*, 1-4.

(41) Zhang, Y.; Hoon, M.A.; Chandrashekar, J.; Mueller, K.L.; Cook, B.; Wu, D.; Zuker, C.S.; Ryba, N.J.P. *Cell* **2003**, *112*, 293-301.

(42) Perez, C.A.; Huang, L.; Rong, M.; Kozak, J.A.; Preuss, A.K.; Zhang, H.; Max, M.; Margolskee, R.F. *Nat. Neurosci.* **2002**, *5*, 1169-1176.

(43) Caterina, M.J.; Schumacher, M.A.; Tominaga, M.; Rosen, T.; Levine, J.D.; Julius, D. *Nature* **1997**, *389*, 816-824.

(44) Caterina, M.J.; Rosen, T.; Tominaga, M.; Brake, A.J.; Julius, D. *Nature* **1999**, *398*, 436-441.

(45) McKerny, D.D.; Neuhausser, W.M.; Julius, D. *Nature* **2002**, *416*, 52-58.

(46) Viana, F.; De la Pena, E.; Belmonte, C. *Nature Neuroscience* **2002**, *5(3)*, 254-260.

(47) Bryant, B.; Mezine, I. In: *Chemsitry of Taste* (Given P., Paredes D., eds.), ACS Symposium Series 825, American Chemical Society, Washington DC, 2002, 202-212.

(48) Bryant, B.P.; Mezine, I. *Brain Res.* **1999**, *842*(2), 452-460.

(49) Woodroof, J.G.; Philips, G.F. In: *Beverages; Carbonated and Non-Carbonated.* Avi Publishing Co. Westport, CT, **1981**.

(50) Green, B.G. *Chem. Senses* **1992**, *17*, 435-450.

(51) McEvoy, S. In: *Chemical senses day XIV abstracts*, Santa Rosa, CA, **1998**.

(52) Graber, M.; Kelleher, S. *Am. J. Med.* **1988**, *84*, 979-980.

(53) Simmons, C.T.; Dessirier, J.-M.; Carstens, M.I.; O'Mahony, M.; Carstens, E. *J. Neurosci.* **1999**, *19*(18), 8134-8144.

(54) Dessirier, J.-M.; Simons, Ch.T.; Carstens, M.I.; O'Mahony; M.; Carstens, E. *Chem. Senses* **2000**, *25*, 277-284.

(55) Leinonen, J.; Parkkila, S.; Launisto, K.; Koivunen, P.; Rajaniemi, H. *J. Histo. Cyto.* **2001**, *498*, 657-662.

(56) Henkin, R.I.; Martin, B.M.; Agarwal, R.P. *Am. J. Med. Sci.* **1999**, *318*, 380-391.

(57) Bathe-Smith, E.C.; Swain, T. In: *Comparative Biochemistry* (Mason, H.S., Florkin, A.M.; Eds.), Academic Press, New York, N, 162; pp 755-809.

(58) Hagerman, A.E.; Butler, L.G. *J. Biol. Chem.* **1981**, *256*, 4494-4497.

(59) Lyman, B.J.; Green, B.G. *Chem. Senses* **1990**, *15*, 151-164.

(60) Naish, M.; Clifford, M.N.; Birch, G.G. *J. Sci. Food Agric.* **1993**, *61*, 57-64.

(61) Smith, A.K.; June, H.; Noble, A.C. *Food Qual. Pref.* **1996**, *7*, 161-166.

(62) Ishikawa, T.; Noble, A.C. *Food Qual. Pref.* **1995**, *6*, 27-64.

(63) Noble, A.C. In: *Chemistry of taste – mechanism, behaviours, and mimics* (P. Given, D. Paredes, Eds.) ACS Symposium Series 825, American Chemical Society, Washington, DC, USA, 2002, p. 192-201.

(64) Peleg, H.; Gacon, K.; Noble, A.C. *F. Sci. Food Agric.* **1999**, *79*, 1123-1128.

(65) Courregelongue, A.; Schlich, P.; Noble, A.C. *Food Qual. Pref.* **1999**, *10*, 273-279.

(66) Gilbertson, T.A. *Current. Opt. Neurobiol.* **1998**, *8*, 447-452.

(67) Gilbertson, T.A.; Kim, I. In: *Chemistry of taste – mechanism, behaviours, and mimics* (P. Given, D. Paredes, Eds.) ACS Symposium Series 825, American Chemical Society, Washington, DC, USA, 2002, p. 180-191.

(68) Hamosh, M.; Burns, W.A. *Lab. Invest.* **1977**, *37*, 603-608.

(69) Fukawatari, T.; Kawada, T.; Tsuruta, M.; Hiraoka, T.; Iwanaga, T.; Sugimoto, E.; Fushiki, T. *FEBS Lett.* **1997**, *414*, 461-464.

(70) Kock, K.; Bläker, M.; Schmale, H. *Cell. Tissue Res.* **1992**, *267*, 313-320.

(71) Gilbertson, T.A.; Fontenot, D.T.; Liu, L.; Zhang, H.; Monroe, W.T. *Am. J. Physiol.* **1997**, *272*, C1203-C1210.

(72) Mattes, R.D. *Am. J. Clin. Nutr.* **1996**, *63*, 911-917.

(73) Bacon, A.W.; Miles, J.S; Schiffman, S.S. *Physiol. Behav.* **1994**, *55*, 603-606.

(74) Bartoshuk, L.M.; Duffy, V.B.; Miller, I.J. *Physiol. Behav.* **1994**, *56*, 1155-1171.

(75) Tepper, B.J.; Nurse, R.J. *Physiol. Behav.* **1997**, *61*, 949-954.

Molecular Physiology and Taste Coding

Chapter 2

Insights into Taste Transduction and Coding from Molecular, Biochemical, and Transgenic Studies

Robert F. Margolskee

Department of Physiology and Biophysics, Howard Hughes Medical Institute, 1425 Madison Avenue, Box 1677, The Mount Sinai School of Medicine of New York University, New York, NY 10029

This review describes studies in taste transduction from my group during the past decade. We have used biochemistry, molecular cloning and bioinformatics to identify and characterize the receptors, G proteins, ion channels and second messenger effector enzymes that serve as signaling elements in taste receptor cells. Gustducin is a heterotrimeric G protein selectively expressed in a subset of taste receptor cells. Knockout mice lacking gustducin's α subunit display markedly reduced behavioral and nerve responses to both bitter and sweet compounds, implicating this G protein in transduction of both bitter and sweet compounds. Gustducin's α subunit activates a phosphodiesterase, while gustducin's βγ subunits (β3γ13) activate phospholipase C. Trpm5, a taste cell-expressed transient receptor potential channel, has been implicated in bitter responses. T1r3, a novel taste receptor, is the *sac* gene product and a component of a sweet-responsive taste receptor.

Introduction

The sensation of taste is initiated by the interaction of tastants with receptors and ion channels in the apical microvilli of taste receptor cells. This fundamental sense enables organisms to avoid toxins (e.g. bitter plant alkaloids) and find nutrients (e.g. sweet carbohydrates). Psychophysics argues that human taste is comprised of five distinct qualities (sweet, sour, bitter, salty and *umami* (the taste of glutamate)), which correlate well with taste responses of animals monitored by behavioral and nerve recording. Taste receptor cells are specialized epithelial cells with many neuronal properties including the ability to depolarize and form synapses. Taste receptor cells are typically clustered in groups of ~100 within taste buds and synapse via their basolateral aspects with afferent taste nerves.

How the vertebrate taste cell responds to a given tastant (*signal detection* and *signal transduction*), and how this information is coded and processed (*sensory coding*) have been the key questions guiding my research for the past decade. In this review I will summarize results obtained in my laboratory during the past ten years by our use of molecular, cellular and whole animal techniques to identify the components of the taste receptor cell and characterize their roles in specifying the peripheral and central taste code.

The Elements of Taste Transduction

Gustducin: a Taste Selective G Protein

Gustducin is a heterotrimeric G protein that is selectively expressed in ~25% of taste receptor cells (*1,2*). Gustducin's α subunit (α-gustducin) is closely related to those of the transducins (80% identical, 90% similar) (*1*), suggesting that α-gustducin might regulate taste phosphodiesterase (PDE) during taste transduction – analogously to the role of the α-transducins in phototransduction. Indeed, PDEs from bovine taste tissue can be activated in vitro by α-gustducin and α-transducin (*3*; Bakre et al., submitted). That PDE activation by α-gustducin may be involved in taste transduction is supported by the observation that several bitter compounds elicit a decrease in cyclic nucleotide levels in taste tissue which can be selectively blocked by antibodies against α-gustducin (*4*). Using native taste receptors from bovine taste tissue we have shown that quinine, naringin, denatonium, strychnine, atropine and several other bitter compounds activate gustducin (Figure 1A) (*5,6*). The Buck (*7*) and Ryba/Zuker (*8*) groups have used molecular cloning to identify the

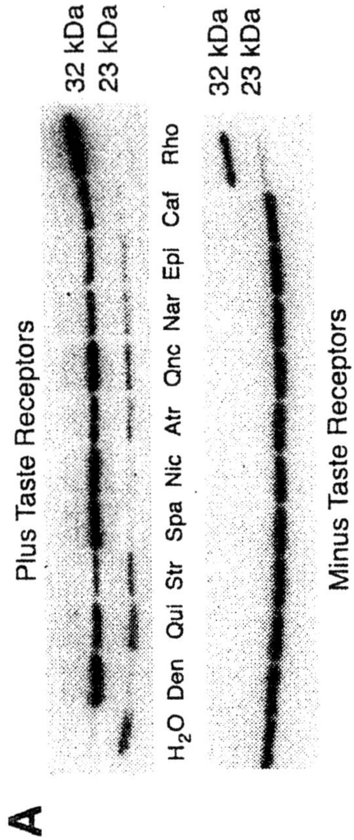
A
Plus Taste Receptors
32 kDa
23 kDa
H$_2$O Den Qui Str Spa Nic Atr Qnc Nar Epi Caf Rho
32 kDa
23 kDa
Minus Taste Receptors

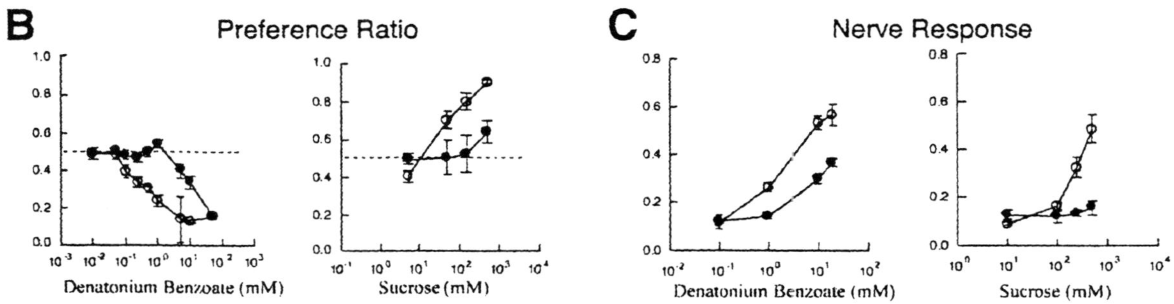

Figure 1. α-Gustducin's role in taste transduction. ***A.*** *Trypsin assays show that in the presence of native taste receptors G protein α-subunits such as α-gustducin or α-transducin (shown) can be activated by several bitter compounds (note the 32 kDa fragment diagnostic of α-transducin activation).* ***B.*** *Mean preference ratios from 48 hr two-bottle preference tests (consumption of tastant vs. water) show that in comparison to wildtype mice (open circles) α-gustducin null mice (filled circles) require higher concentrations of denatonium benzoate (bitter) or sucrose (sweet) to show avoidance or preference, respectively.* ***C.*** *Summated chorda tympani nerve responses show that in comparison to wildtype mice (open circles) α-gustducin null mice (filled circles) display diminished nerve responses to denatonium benzoate or sucrose. Adapted with permission from references 5 and 11. Copyright 1998, 1996.*

T2r/Trb family of taste receptors. A few members of this family of receptors have been shown to respond to various bitter compounds and to selectively couple to gustducin over other G proteins (*9,10*), suggesting that many, perhaps even all, of these receptors are bitter-responsive.

In vivo results from α-gustducin knockout mice have demonstrated clearly that gustducin plays a key role in taste receptor cell responses to bitter and sweet compounds (*11*). α-Gustducin knockout mice displayed markedly reduced behavioral and nerve responses to the bitter compounds denatonium benzoate and quinine sulfate, and to the sweet compounds sucrose and SC45647 (Figure 1BC). More extensive analysis of taste responses of α-gustducin knockout mice, α-transducin knockouts and α-transducin/α-gustducin double knockouts have shown that α-gustducin is essential to responses to most bitter, sweet and umami compounds, while α-transducin is only involved in taste responses to umami (Danilova, V., He, W., Glendinning, J., Hellekant, G., Damak, S., Margolskee, R. F., unpublished). Transgenic expression of a dominant-negative mutant of α-gustducin inhibited the residual sweet and bitter responsiveness of α-gustducin knockout mice, arguing that another taste-expressed G protein coupled to taste receptors mediated this effect (*12*). Analysis of α-transducin/α-gustducin double knockouts indicates that α-transducin does not mediate these residual responses.

Gγ13: Gustducin's γ-subunit

We used single cell reverse transcription-polymerase chain reaction (RT-PCR) and differential screening to clone cDNAs specific to α-gustducin-positive taste receptor cells (*13*). Of 40,000 plaques screened, 60 clones were differentially positive when probed with cDNA probes from α-gustducin-positive vs. α-gustducin-negative cells. Two of these clones contained an open reading frame of 201 base pairs, predicted to encode a novel 67 amino acid long G protein γ-subunit, named Gγ13. Gγ13 is only distantly related to the known G protein γ-subunits (25-33% identity). By northern blot we detected a Gγ13 transcript of 0.5 kb in brain, retina, olfactory epithelium, and to a lesser extent, in stomach and testis (Figure 2A). The greatest labeling was seen in olfactory epithelium and cerebellum, followed by retina. In situ hybridization with taste bud-containing tissue showed that Gγ13 mRNA was selectively expressed in taste receptor cells, but absent from the surrounding lingual epithelium, muscle or connective tissue (*13*). We used "expression profiling" with taste bud-containing tissue and control non-gustatory lingual epithelia to determine that α-gustducin, Gβ3 and Gγ13 were only expressed in taste bud-containing tissue, while Gβ1 was expressed in both gustatory and non-gustatory lingual epithelia (Figure 2B). Profiling the pattern of expression of individual taste receptor cells showed that 19 of 19 cells that expressed α-gustducin also expressed Gγ13 and β3 (Figure 2B).

We used three different trypsin protection assays to monitor interactions of Gγ13 with other G protein subunits (Figure 2C). In this way we found that Gγ13 interacted in vitro with α-gustducin. Gγ13 also interacted with Gβ1 and Gβ4 to form dimers (Figure 2C). Because of difficulties in expressing Gβ3 it was not possible to use the trypsin assay to assess interactions between Gγ13 and Gβ3.

In a third type of trypsin assay we determined that Gβ1γ13 dimers enhanced the activation of α-gustducin by taste receptor-containing membranes stimulated by the bitter compound denatonium (Figure 2C). The activation of α-gustducin required: 1. taste receptor-containing membranes; 2. the bitter compound denatonium and; 3. a Gγ13-containing βγ dimer. Hence, we conclude that α-gustducin, Gβ1 and Gγ13 can associate with each other to form a functional heterotrimeric G protein capable of interacting with denatonium-responsive taste receptors.

To determine if Gγ13 does indeed function in taste transduction we carried out quench-flow experiments with taste tissue. The addition of bitter denatonium benzoate to murine taste tissue induced the generation of IP_3 to slightly more than twice the basal level (Figure 2D). When the taste tissue was preincubated with antibodies to Gγ13 the addition of denatonium did not increase IP_3 levels appreciably (Figure 2D). Preincubation with antibodies against Gγ1 or Gγ3 did not reduce denatonium-stimulated generation of IP_3 (data not shown). From these results we conclude that βγ subunit pairs containing Gγ13 mediate the denatonium-responsive activation of taste tissue PLCβ2 to generate IP_3.

This clarifies the previously puzzling observation that many bitter compounds thought to be transduced by gustducin lead to the generation of IP_3 (*14,15*), yet antibodies directed against α-gustducin did not block this response, although they did block the gustducin-mediated decrease in taste receptor cell cyclic nucleotides (*4*). Based on our study of Gγ13 (*13*) and work from the Breer (*16,17*) and Spielman (*4,14,15*) laboratories we can conclude that Gγ13, Gβ3 and PLCβ2, mediate the taste receptor cell's IP_3 response to bitter compounds. Thus, heterotrimeric gustducin mediates two responses in taste receptor cells: a decrease in cyclic nucleotides via α-gustducin activation of phosphodiesterase and a rise in IP_3 via βγ-gustducin (Gβ3γ13) activation of PLCβ2.

Trpm5: a Taste-Selective Store Operated Channel

From the same single taste cell RT-PCR-differential screen which we had used to identify Gγ13 (*13*) we also identified a partial clone that was selectively expressed in α-gustducin-positive taste cells. The full-length clone contained an open reading frame of 4077 bp predicted to encode a protein of 1158 amino acids. The encoded protein was identified as a novel member of the TRP (Transient Receptor Potential) family of ion channels. This TRP channel was identified independently by others and named Mtr1 (18-20) and subsequently renamed Trpm5 (*21*).

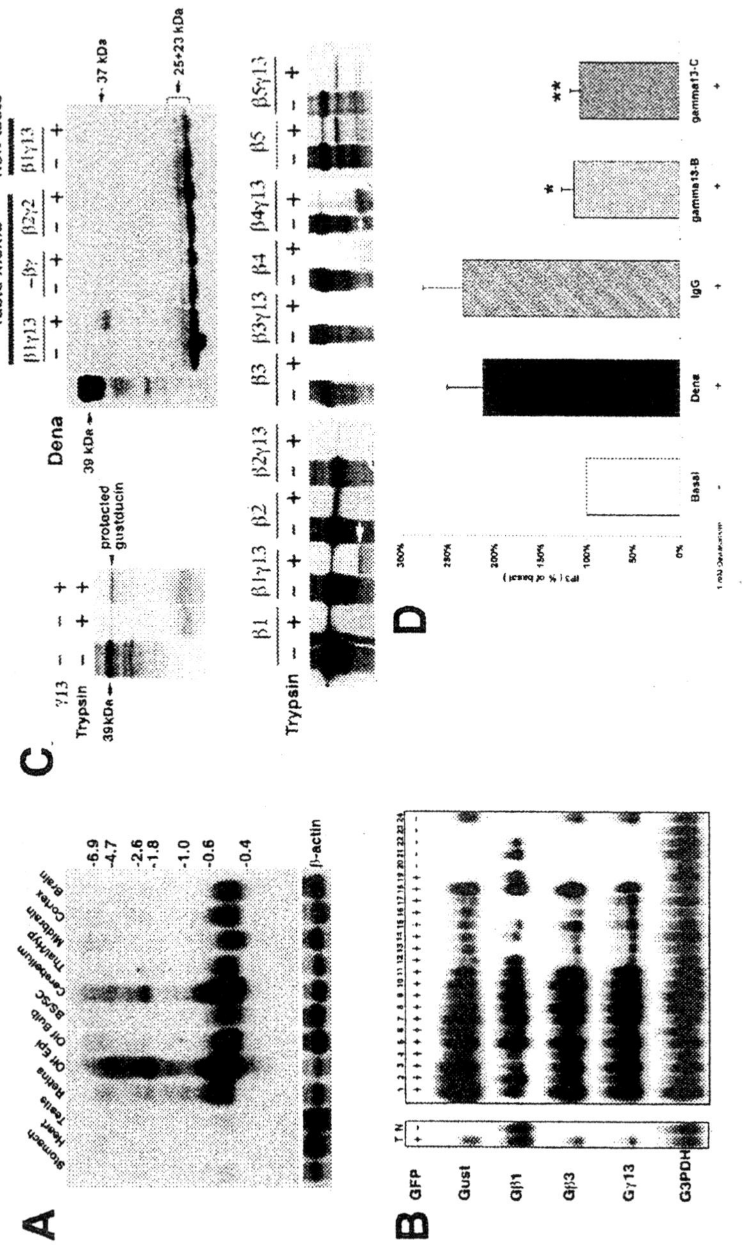
A
Stomach
Heart
Testis
Retina
Olf Epi
Olf Bulb
BS/SC
Cerebellum
Thal/Hyp
Midbrain
Cortex
Brain
-6.9
-4.7
-2.6
-1.8
-1.0
-0.6
-0.4
β-actin
B
GFP
Gust
Gβ1
Gβ3
Gγ13
G3PDH
T N
C
γ13
Trypsin
39kDa
protected gustducin
Dena
39 kDa
Taste memb
non-taste
β1γ13
-βγ
β2γ2
β1γ13
37 kDa
25+23 kDa
β1
β1γ13
β2
β2γ13
β3
β3γ13
β4
β4γ13
β5
β5γ13
D
Basal
Dena
IgG
gamma13-B
gamma13-C

Figure 2. Gγ13's role in taste transduction. ***A.*** *The northern blot shows that Gγ13 is expressed at high levels in mouse brain, retina and olfactory epithelium. The size of the RNA markers (in kilobases) is indicated in the right margin. Abbreviations: olf epi, olfactory epithelium; olf bulb, olfactory bulb; BS/SC, brain stem & spinal cord; thal/hyp, thalamus & hypothalamus.* ***B.*** *Expression profiling shows that Gγ13 is selectively expressed in taste cells along with α-gustducin and Gβ3. Left panel: Southern hybridization to RT-PCR products from mouse taste tissue (T) and control non-taste lingual tissue (N). Right panel: Southern hybridization to RT-PCR products from 24 individually amplified taste receptor cells. 3'-region probes from α-gustducin (Gust), Gβ1, Gβ3, Gγ13, and glyceraldehyde 3-phosphate dehydrogenase (G3PDH) were used to probe the blots.* ***C.*** *Tryptic digestion assays to monitor interactions between Gγ13, α-gustducin and Gβ subunits. Left panel: α-Gustducin is protected (arrow) from tryptic digestion in the presence of Gγ13 indicating an interaction between α-gustducin and Gγ13. Bottom panel: Gβ1 and Gβ4 are protected (arrows) from tryptic digestion in the presence of Gγ13 indicating an interaction between these Gβ-subunits and Gγ13. Right panel: α-Gustducin is activated by denatonium benzoate (Dena) in the presence of taste receptor-containing membranes and Gβ1γ13 (note appearance of the 37 kDa fragement).* ***D.*** *Denatonium-induced IP_3 production in mouse taste tissue is suppressed by anti-Gγ13 antibodies, implicating Gγ13 in mediating this response. (*) Significantly different compared to Dena ($p<0.05$) and Dena+IgG ($p<0.005$); (**) significantly different compared to Dena ($p< 0.01$) and Dena+IgG ($p< 0.001$). Adapted witih permission from reference 13. Copyright 1999.*

In northern blots a transcript of 4.5 kb was found in taste tissue, with no detectable expression in non-taste tissue (Figure 3A). Lower expression was also detected in stomach and small intestine, and very weak expression was seen in uterus and testis (Figure 3A). By in situ hybridization it was determined that Trpm5 mRNA was present in taste receptor cells in circumvallate (Figure 3BC) as well as in foliate and fungiform papillae (data not shown), but not in the surrounding non-gustatory epithelia. The general pattern of Trpm5 expression in taste buds was comparable to that of α-gustducin and Gγ13. "Expression profiling" of a panel of taste receptor cells showed that Trpm5 is co-expressed with α-gustducin, Gγ13 and PLCβ2 (Figure 3D) (this pattern was confirmed by immunohistochemical staining (*22*)).

To determine if Trpm5 functions as an ion channel, we expressed it in *Xenopus* oocytes (Figure 3EF) and CHO cells (data not shown). In voltage clamp experiments we utilized the oocyte's endogenous calcium-activated chloride conductance (ICl_{Ca}) as a reporter to detect changes in calcium conductance due to expression of Trpm5. Trpm5 cRNA was injected into *Xenopus* oocytes and voltage clamp recordings were performed two days later. We incubated the oocytes for 2 h before the recording in 2 μM thapsigargin, an irreversible inhibitor of the sarco-endoplasmic reticulum Ca^{2+}-ATPase, to show that Ca^{2+} flux through Trpm5 was activated by depletion of Ca^{2+} stores (Figure 3E). To confirm that these effects depended on expression of Trpm5 we co-injected oocytes with Trpm5 and a dominant-negative construct ($TRPM5^{DN}$) comprised of transmembrane segments 5 and 6 and the C-terminal region of Trpm5. Co-injection of Trpm5 and $TRPM5^{DN}$ cRNAs abolished the Trpm5-dependent activation of ICl_{Ca} confirming that Trpm5 does act as a store operated channel.

Ca^{2+}-imaging of oocytes demonstrated increased calcium permeability in oocytes injected with Trpm5 cRNA vs. those injected with water (Figure 3F). When thapsigargin was used to deplete Ca^{2+} stores, the Trpm5-injected oocytes displayed a greater influx of Ca^{2+} than did the water injected controls (Figure 3F). The average fluorescent intensity of the Trpm5-injected oocytes was much greater than that of the water-injected controls and was most noticeable in the presence of thapsigargin. These results corroborate the observations obtained from electrophysiological recordings in Trpm5-transfected CHO cells and Trpm5 cRNA-injected *Xenopus* oocytes (Figure 3E) and are consistent with Trpm5's potential function as a calcium channel activated by depletion of Ca^{2+} stores.

We collaborated with Kinnamon to look at store operated channels in vivo in α-gustducin-positive/Trpm5-positive taste receptor cells which had been marked by transgenic expression of GFP (green fluorescent protein) (*13*). The GFP-positive taste receptor cells responded to thapsigargin depletion of Ca^{2+} stores or stimulation with the bitter compound denatonium benzoate with an influx of Ca^{2+} (*23*). Trpm5 may act as a store operated channel to promote calcium entry in taste receptor cells in response to bitter and/or sweet compounds, which are known to elicit Ca^{2+} influx as well as Ca^{2+} release from internal stores (*24-26*). Influx of Ca^{2+} into taste receptor cells via Trpm5 may increase the likelihood or amount of neurotransmitter release by the taste cell, and hence the strength of the gustatory signal.

Sweet-responsive receptors

It has been known for ~30 years that inbred strains of mice differ markedly in their ingestive responses to saccharin and sweeteners (*27,28*). The *sac* locus is the primary genetic determinant of responsiveness of mice to sweeteners (*29*). To identify the *sac* gene we used bioinformatics to order into a contiguous stretch of DNA all predicted genes present within a region of ~1 million bp of the sequenced (but at the time, unfinished) human genome syntenous to the *sac* region of mouse chromosome 4 (Figure 4A) (*30*). Within this region of human chromosome 1p36 we identified human *TAS1R3*, a previously unknown GPCR and the only GPCR in this region, as the most likely candidate for *sac*. We found that mouse *Tas1r3* maps to the terminal portion of mouse chromosme 4p within 20,000 bp of D18346, the marker most tightly linked to *sac* (Figure 4A).

TAS1R3 and *Tas1r3* encode T1R3 and T1r3, respectively, GPCRs most closely related to the T1r1 and T1r2 taste receptors that function as components of umami- and sweet-responsive receptors, respectively (*31-33*). T1R3 is predicted to encode a protein of 843 amino acids with seven transmembrane helices and a large 558 amino acid long extracellular domain; T1r3 is predicted to be 858 amino acids long with a 573 amino acid long extracellular domain. By northern blot we determined that a 7.2 kb *Tas1r3* mRNA was expressed selectively in taste, but not in control lingual tissue devoid of taste buds (non-taste) or in any of the several other tissues examined (Figure 4B). A somewhat larger (~7.8 kb) mRNA species was expressed at moderate levels in testis, and at very low levels in brain. A smaller (~6.7 kb) mRNA species was expressed at very low levels in thymus. By in situ hybridization we showed that *Tas1r3* mRNA was selectively expressed in taste receptor cells in circumvallate, foliate (Figure 4C) and fungiform (data not shown) papillae, but absent from the surrounding lingual epithelium, muscle or connective tissue. Sense probe controls showed no non-specific hybridization to lingual tissue (data not shown). We examined the nucleotide and inferred amino acid sequence of T1r3 from taster and non-taster strains of mice looking for allelic differences that might explain their phenotypic differences. All four non-taster strains (DBA/2, 129/Svev, BALB/c and C3H/HeJ) examined shared the same allele with identical nucleotide sequence despite the fact that their most recent common ancestors date back to the early 1900s or earlier (Figure 4D).

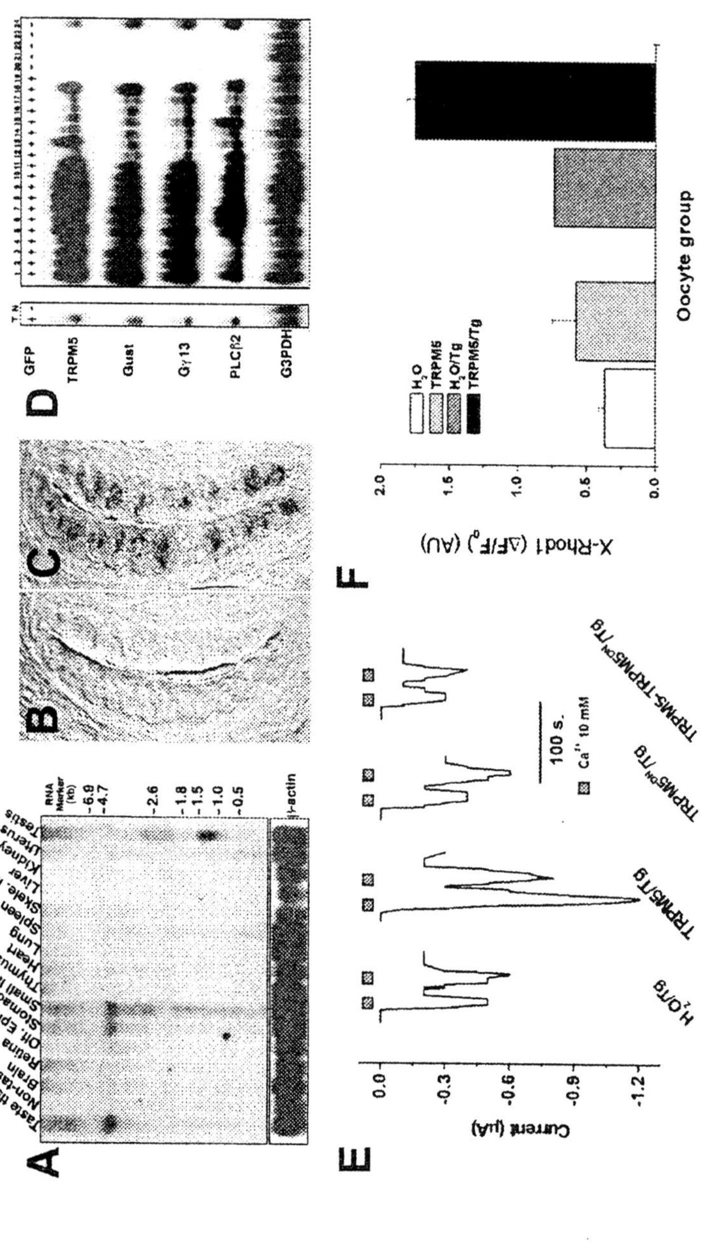
A
Taste tissue
Non-taste
Brain
Retina
Olf. Epi.
Stomach
Small Int.
Thymus
Heart
Lung
Spleen
Skele. Mus.
Liver
Kidney
Uterus
Testis
RNA Marker (kb)
-6.9
-4.7
-2.6
-1.8
-1.5
-1.0
-0.5
β-actin
B
C
D
GFP
TRPM5
Gust
Gγ13
PLCβ2
G3PDH
T N
E
Current (μA)
0.0
-0.3
-0.6
-0.9
-1.2
100 s.
Ca²⁺ 10 mM
H₂O/Tg
TRPM5/Tg
TRPM5ᴰᴺ/Tg
TRPM5-TRPM5ᴰᴺ/Tg
F
X-Rhod1 (ΔF/F₀) (AU)
2.0
1.5
1.0
0.5
0.0
H₂O
TRPM5
H₂O/Tg
TRPM5/Tg
Oocyte group

Figure 3. Trpm5's role in taste transduction. ***A.*** *Northern blot shows restricted expression of Trpm5 mRNA, with highest levels in taste tissue, stomach and intestine.* ***B. & C.*** *In situ hybridization to taste bud-containing circumvallate papillae using TRPM5 sense (B) and antisense (C) probes shows expression of Trpm5 mRNA in a majority of the taste buds.* ***D.*** *Single cell gene expression profiling shows co-expression of Trpm5 in taste receptor cells with other signaling molecules such as α-gustducin (GUST), Gγ13 and PLCβ2.* ***E.*** *Two electrode voltage clamp studies of thapsigargin (Tg) treated Xenopus oocytes heterologously expressing Trpm5 (TRPM5/Tg) vs. control water-injected (H_2O/Tg) oocytes show strong enhancement of the endogenous Ca^{2+}–sensitive Cl^- reporter current (ICl_{Ca}). This effect was abolished when Trpm5 cRNA was co-injected with cRNA from a dominant negative form of Trpm5 ($TRPM5^{DN}$) (see TRPM5-$TRPM5^{DN}$/Tg trace).* ***F.*** *Calcium imaging of oocytes loaded with the Ca^{2+}-sensitive dye X-rhod-1 AM shows an increase in Ca^{2+} influx in those oocytes that were injected with Trpm5 cRNA. This effect was especially pronounced in oocytes that had been pre-treated with thapsigargin (Tg). Adapted with permission from reference 22. Copyright 2002.*

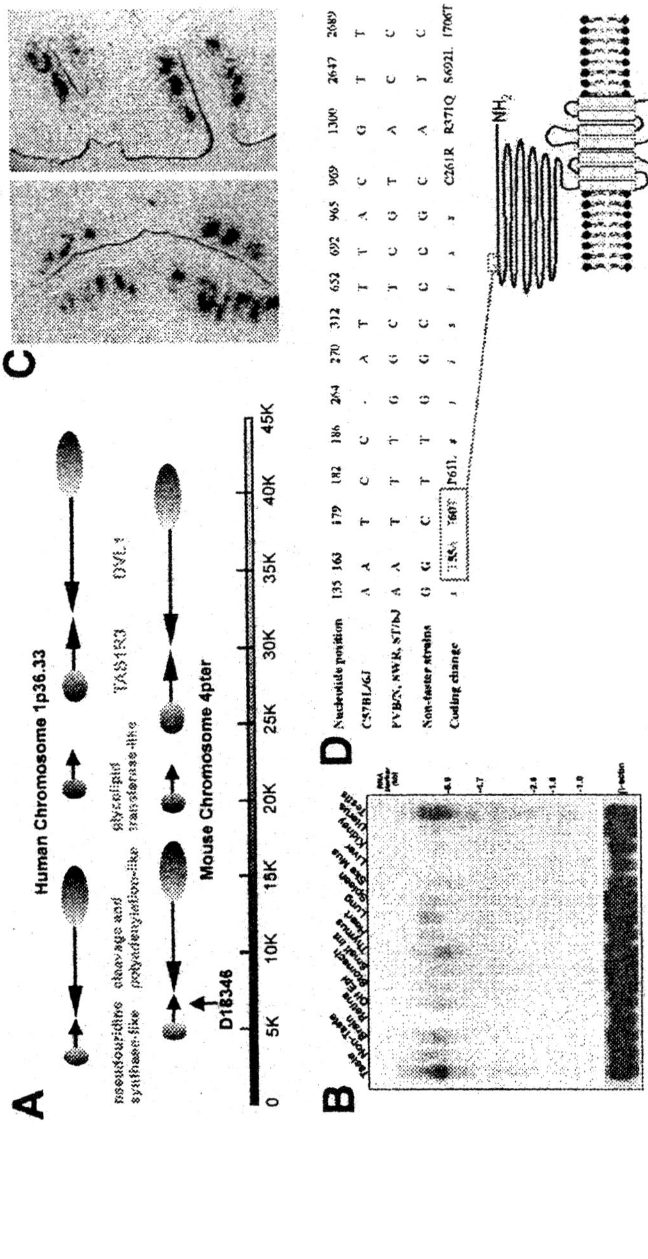
A
Human Chromosome 1p36.33
pseudouridine synthase-like
cleavage and polyadenylation-like
glycolipid transferase-like
TAS1R3
DVL1
D18346
Mouse Chromosome 4pter
0
5K
10K
15K
20K
25K
30K
35K
40K
45K
B
C
D
Nucleotide position
C57BL/6J
Non-taster strains
Coding change
NH2
COOH
T1R3

*Figure 4. T1r3's role in taste transduction.. **A.** Candidate genes in the region of the sac locus on mouse chromosome 4 (4pter) and the syntenic region of human chromosome 1 (1p36.33). Tas1r3 encodes T1r3, a novel GPCR. The region displayed corresponds to ~45,000 bp. **B.** The northern blot shows that a 7.2 kb T1r3 transcript was detected only in the taste tissue, and a slightly larger transcript was detected in testis. The size of the RNA markers (in kilobases) is indicated in the right margin. **C.** In situ hybridization of T1r3 antisense probe to taste bud-containing circumvallate (Left panel) and foliate (Right panel) papillae shows expression of T1r3 mRNA in the majority of the taste buds. **D.** Upper panel: T1r3 allelic differences were determined in eight inbred mouse strains. All non-taster strains showed identical sequences. In the bottom row the amino acid immediately after the position number is always from the non-tasters, while the amino acid immediately before the position number is from whichever tasters differed at that position from the non-tasters. The two columns in bold represent positions where all tasters differed from all non-tasters and where the differences in nucleotide sequence result in amino acid substitutions. Nucleotide differences that do not alter the encoded amino acid are indicated as s: silent. Nucleotide differences within introns are indicated as i: intron. Lower panel: The schematic diagram shows that the T55A and I60T polymorphisms that distinguished all tasters from all non-tasters are located within the extracellular amino terminal domain of T1r3. Adapted with permission from reference 30. Copyright 2001.*

The four taster strains fell into two allelic classes (C57BL/6J vs. SWR, FVB/N and ST/bj). Fifteen positions were polymorphic at the nucleotide level, seven of these positions are predicted to lead to coding differences (the other nucleotide differences were in introns or were "silent" alternate codon changes) (Figure 4D). There were only two coding differences that distinguished all tasters from all non-tasters: **T55A** and **I60T** (single letter amino acid code, with the non-taster amino acid listed last). The **I60T** change is particularly intriguing difference as it is predicted to introduce a novel N-linked glycosylation site in the extracellular amino acid terminal domain of T1r3 that, according to models of T1r3's structure, is predicted to interfere with dimerization. (30). Transgenic expression in non-taster mice of T1r3 from a taster strain manifests the taster phenotype (31; Rong, M., He, W., Damak, S., Margolskee, R.F., unpublished.), confirming that T1r3 is sac.

Bitter and Sweet Taste Transduction Pathways

Bitter

Gustducin heterotrimers that have been activated by bitter-stimulated T2r/Trb receptors mediate two responses in taste receptor cells: a decrease in cyclic nucleotides (cNMPs) via α-gustducin and a rise in IP_3 and diacyl glycerol (DAG) via βγ-gustducin. The subsequent steps in the α-gustducin-PDE-cNMP pathway are presently uncertain (reviewed in *34*): decreased cNMPs may act on protein kinases which in turn may regulate taste receptor cell ion channel activity, or cNMP levels may regulate directly the activity of cNMP-gated (*35*) and cNMP-inhibited (*36*) ion channels expressed in taste receptor cells. The subsequent steps in the βγ-gustducin-PLC-IP_3/DAG pathway are apparently activation of IP_3 receptors (type 3 IP_3 receptors have recently been shown to be co-expressed in TRCs with Gγ13 and PLCβ2 (*37,38*)) and release of Ca^{2+} from internal stores followed by neurotransmitter release (*26*). These pathways are diagrammed in Figure 5.

Sweet

Based on biochemical and electrophysiological studies of taste receptor cells (*26,39-46*) two models for sweet transduction have been proposed. First, a GPCR-G_s-cAMP pathway – sucrose and other sugars lead to activation of G_S via one or more coupled GPCRs (presumably T1r heterodimers); receptor-activated $G\alpha_S$ activates adenylyl cyclase (AC) to generate cAMP; cAMP may act directly to cause cation influx through cNMP gated channels, or act indirectly to activate

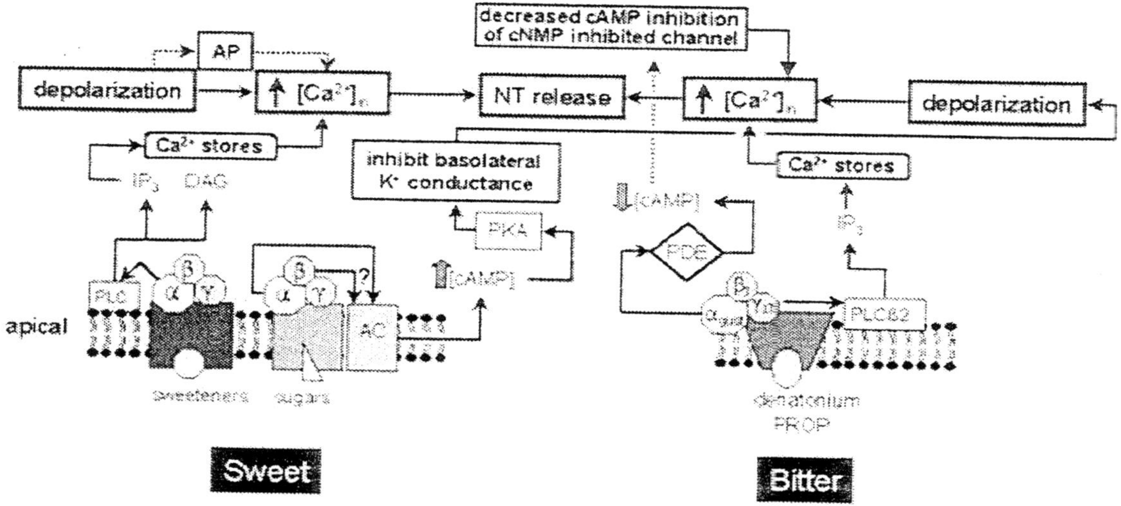

Figure 5. Proposed transduction mechanisms in vertebrate taste receptor cells underlying bitter and sweet taste qualities. All transduction pathways are proposed to converge on common elements which mediate a rise in intracellular Ca^{2+} followed by neurotransmitter (NT) release. Adapted with permission from reference 47. Copyright 2002.

protein kinase A which phosphorylates a basolateral K^+ channel, leading to closure of the channel, depolarization of the taste cell, voltage-dependent Ca^{++} influx and neurotransmitter release. Second, a GPCR-G_q/$G\beta\gamma$-IP_3 pathway - artificial sweeteners presumably bind to and activate one or more GPCRs (T1r heterodimers?) coupled to PLCβ2 by either the α subunit of G_q or by Gβγ subunits; activated $G\alpha_q$ or released Gβγ activates PLCβ2 to generate IP_3 and DAG; IP_3 and DAG elicit Ca^{++} release from internal stores, leading to depolarization of the TRC and neurotransmitter release. These two pathways (diagrammed in Figure 5) coexist in the same taste receptor cells: sweet-responsive taste receptor cells from rat circumvallate papillae had an influx of Ca^{2+} in response to sucrose, while the artificial sweeteners saccharin and SC45647 elevated Ca^{++} via release from internal stores (*26*). These sweet-responsive taste receptor cells did <u>not</u> respond to any bitter stimuli (*26*). Now that the sweet-responsive T1r receptors have been cloned and expressed it should be possible to definitively test these various models of sweet transduction. It is presently unclear how these receptors could selectively mediate cAMP responses to sugars and IP_3 responses to artificial sweeteners.

References

1. McLaughlin, S.K.; McKinnon, P.J.; Margolskee, R.F. *Nature* **1992**, *357*, 563-569.
2. Boughter Jr, J.D.; Pumplin, D.W.; Yu, C.; Christy, R.C.; Smith, D.V. *J. Neurosci.* **1997**, *17*, 2852-2858.
3. Ruiz-Avila, L.; McLaughlin, S.K.; Wildman, D.; McKinnon, P.J.; Robichon, A.; Spickofsky, N.; Margolskee, R.F. *Nature* **1995,** *376*, 80-85.
4. Yan, W.; Sunavala, G.; Rosenzweig, S.; Dasso, M.; Brand, J.G.; Spielman, A.I. *Am. J. Physiol. Cell. Physiol.* **2001,** *280,* C742-C751.
5. Ming, D.; Ruiz-Avila, L.; Margolskee, R.F. *Proc. Natl. Acad. Sci. USA* **1998,** *95*, 8933-8938.
6. Ruiz-Avila, L.; Ming, D.; Margolskee, R.F. *Chem. Senses* **2000,** *25*, 361-368.
7. Matsunami, H.; Montmayeur, J-P; Buck, L.B. *Nature* **2000,** *404*, 601-604.
8. Adler, E.; Hoon, M.A.; Mueller, K.L.; Chandrashekar, J.; Ryba, N.J.P.; Zuker, C.S. *Cell* **2000,** *10*, 693-702.
9. Chandrashekar, J.; Mueller, K.L.; Hoon, M.A.; Adler, E.; Feng, L.; Guo, W.; Zuker, C.S.; Ryba, N.J. *Cell* **2000,** *100*, 703-711.
10. Bufe, B.; Hofmann, T.; Krautwurst, D.; Raguse, J.D.; Meyerhof, W. *Nat. Genet.* **2002**, *32(3)*, 397-401.
11. Wong, G.T.; Gannon, K.S.; and Margolskee, R.F. *Nature* **1996,** *381*, 796-800.
12. Ruiz-Avila, L.; Wong, G.T.; Damak, S.; Margolskee, R.F. *Proc. Natl. Acad. Sci., USA* **2001**, *98*, 8868-8873.

13. Huang, L.; Y. Shanker, G.; Dubauskaite, J.; Zheng, J.Z.; Yan, W.; Rosenzweig, S.; Spielman, A.I.; Max, M.; and Margolskee, R.F. *Nature Neurosci.* **1999,** *2,* 1055-1062.
14. Spielman, A.I.; Huque, T.; Nagai, H.; Whitney, G.; Brand, J. *Physiol. Bhvr.* **1994,** *56,* 1149-1155.
15. Spielman, A.I.; Nagai, H.; Sunavala, G.; Dasso, M.; Breer, H.; Boekhoff, I.; Huque, T.; Whitney, G.; Brand, J. *Am. J. Physiol.* **1996**, *270*, C926-C931.
16. Rossler, P.; Kroner, C.; Freitag, J.; Noe, J.; Breer, H. *Eur. J. Cell. Biol.* **1998,** *77,* 253-261.
17. Rossler, P.; Boekhoff, I.; Tareilus, E.; Beck, S.; Breer, H.; Freitag, J. *Chem. Senses* **2000,** *25,* 413-421.
18. Yatsuki, H.; Watanabe, H.; Hattori, M.; Joh, K.; Soejima, H.; Komoda, H.; Xin, Z.; Zhu, X.; Higashimoto, K.; Nishimura, M.; Kuratomi, S.; Sasaki, H.; Sakaki, Y.; Mukai, T. *DNA Res.* **2000,** *7,* 195-206.
19. Enklaar, T.; Esswein, M.; Oswald, M.; Hilbert, K.; Winterpacht, A.; Higgins, M.; Zabel, B.; Prawitt, D. *Genomics* **2000,** *67,* 179-187.
20. Prawitt, D.; Enklaar, T.; Klemm, G.; Gartner, B.; Spangenberg, C.; Winterpacht, A.; Higgins, M.; Pelletier, J.; Zabel, B. *Hum. Mol. Genet.* **2000,** *9,* 203-216.
21. Montell, C.; Birnbaumer, L.; Flockerzi, V.; Bindels, R.J.; Bruford, E.A.; Caterina, M.J.; Clapham, D.E.; Harteneck, C.; Heller, S.; Julius, D.; Kojima, I.; Mori, Y.; Penner, R.; Prawitt, D.; Scharenberg, A.M.; Schultz, G.; Shimizu, N.; Zhu M.X. *Mol. Cell.* **2002,** *9,* 229-231.
22. Perez, C.A.; Huang, L.; Rong M.; Kozak, J.A.; Preuss, A.K.; Max, M.; Margolskee, R.F. *Nature Neurosci.* **2002,** *5,* 1169-1176.
23. Ogura, T.; Margolskee, R.F.; Kinnamon, S.C. *J. Neurophysiol.* **2002,** *87,* 3152-3155.
24. Akabas, M.H.; Dodd, J.; Al-Awqati, Q. *Science* **1988,** *242,* 1047-1050.
25. Orola, C.N.; Yamashita, T.; Harada, N.; Amano, H.; Ohtani, M.; Kumazawa, T. *Acta. Otolaryngol.* **1992,** *112,* 120-127.
26. Bernhardt, S.J.; Naim, M.; Zehavi, U.; Lindemann, B. *J. Physiol.* **1996,** *490,* 325-336.
27. Capretta, P.J. *Psychon. Science* **1970,** *21,* 133-135.
28. Pelz, W.; Whitney, G.; Smith, J.C. *Physiol. Bhvr.* **1973,** *10,* 263-265.
29. Fuller, J.L. *J. Hered.* **1974,** *65,* 33-36.
30. Max, M.; Shanker, Y. G.; Huang, L.; Rong, M.; Liu, Z.; Campagne, F.; Weinstein, H.; Damak, S.; Margolskee, R.F. *Nature Genet.* **2001,** *28,* 58-63.
31. Nelson, G.; Hoon, M.A.; Chandrashekar, J.; Zhang, Y.; Ryba, N.J.P.; Zuker, C.S. *Cell* **2001,** *106,* 381-390.
32. Nelson, G.; Chandrashekar, J.; Hoon, M.A.; Feng, L.; Zhao, G.; Ryba, N.J.; Zuker, C.S. *Nature* **2002,** *416,* 199-202.
33. Li, X.; Staszewski, L.; Xu, H.; Durick, K.; Zoller, M.; Adler, E. *Proc. Natl. Acad. Sci. USA* **2002,** *99,* 4692-4696.
34. Spielman, A.I. *J. Dent. Res.* **1998,** *77,* 539-544.
35. Misaka, T.; Kusakabe, Y.; Emori, Y.; Gonoi, T.; Arai, S.; Abe, K. *J. Biol. Chem.* **1997,** *272,* 22623-22629.

36. Kolesnikov, S.; Margolskee, R.F. *Nature* **1995,** *376*, 85-88.
37. Clapp, T.R.; Stone, L.M.; Margolskee, R.F.; Kinnamon, S.C. *BMC Neurosci.* **2001**, *6*, 2-10.
38. Miyoshi, M.A.; Abe, K.; Emori, Y. *Chem. Senses* **2001**, *26*, 259-265.
39. Striem, B.J.; Pace, U.; Zehavi, U.; Naim, M.; Lancet, D. *Biochem. J.* **1989,** *260*, 121-126.
40. Striem, B.J.; Naim, M.; and Lindemann, B. *Cell. Physiol. and Biochem.* **1991**, *1*, 46-54.
41. Naim, M.; Ronen, T.; Striem, B.J.; Levenson, M.; Zehavi, U. *Comp. Biochem. Physiol.* **1991**, *100B*, 455-458.
42. Tonosaki, K.; Funakoshi, M. *Nature* **1988**, *331*, 354-356.
43. Avenet, P.; Hofmann, F.; Lindemann, B. *Nature* **1988**, *331*, 351-354.
44. Avenet, P.; Hofmann, F.; Lindemann, B. *Comp. Biochem. Biophysiol.* **1988,** *90A*, 681-685.
45. Cummings, T.A.; Powell, J.; Kinnamon, S.C. *J. Neurophys.* **1993**, *70*, 2326-2336.
46. Cummings, T.A.; Daniels C.; Kinnamon, S.C. *J. Neurophys.* **1996**, *75*, 1256-1263.
47. Margolskee, R.F. *J. Biol. Chem.* **2002**, 277, 1-4.

Chapter 3

Identification of Human Bitter Taste Receptors

Bernd Bufe[2], Ellen Schöley-Pohl[1], Dietmar Krautwurst[1], Thomas Hofmann[2], and Wolfgang Meyerhof[1]

[1]German Institute of Hunam Nutrition, Department of Molecular Genetics, Arthur-Scheunert Allee 114–116, 14558 Potsdam-Rehbrücke, Germany
[2]Institute for Food Chemistry, University of Münster, Correnstrasse 45, 48149 Münster, Germany

Increasing the content of chemoprotective compounds in plant food is a potent dietary option for the prevention of diseases and a challenge for the design of novel food. However, the consumer rejects food rich in these bitter tasting phytonutrients thereby promoting their removal during breeding and food processing. The development of bitter masking agents could circumvent this problem, but require a detailed understanding of the interaction of bitter compounds with their receptors on the human tongue, which, however, has not been achieved yet. We use gene cloning, cell based functional expression systems and human psychophysics to identify and characterize human bitter taste receptors. We demonstrate that hTAS2R16 and hTAS2R10 are receptors for various bitter glycosides and the alkaloid strychnine, respectively. Our data facilitate predictions about structural requirements conditional for receptor-agonist interactions and enable us to search for bitter blocking activities.

Introduction

Bitter taste generally is aversive across species including humans and predicts toxicity. It permits organisms to evaluate their food for the presence of potentially harmful substances of numerous chemical classes. Alkaloids, glycosides, rancid fats, hydrolyzed proteins, certain microbial fermentation products and xenobiotics are examples for bitter tasting compounds, which frequently contaminate human food. On the other hand, a slight or moderate bitter taste is expected in some foods and beverages and contribute to their attraction and palatability. Well known examples include the humolones and lupolones in beer, caffeine in coffee or quinine in tonic water. Moreover, some bitter phytonutrients notably exert beneficial effects. Examples include the bitter principles in digestives, the glycoside salicin of willow bark, with antipyretic analgesic action, or certain flavonoids, which are chemoprotective and lower the risk of cardiovascular and cancer diseases (*1*). The design of functional food rich in healthy bitter phytonutrients is a dietary option and a challenge for food industry, because bitter tasting food is generally rejected by the consumer (*2*). The addition of compounds, which mask the bitter taste of healthy phytonutrients, to functional food would increase its acceptance. The design and employment of bitter blocking agents requires a detailed understanding of the interactions of bitter chemicals with their receptors on the human tongue, which is still pending.

Recent progress elucidated important principles of bitter transduction in the taste receptor cells on the tongue (*3*). Bitter taste transduction involves the heterotrimeric G protein α-gustducin/$G\beta_{3}/G\gamma_{13}$ (*4*). α-Gustducin gene-targeted mice show diminished behavioral and nerve responses to a number of bitter chemicals (*5*). Activation of gustducin splits signal transduction into two pathways. The α-subunit stimulates phosphodiesterase activity resulting in reduced cyclic nucleotide levels in taste tissue (*6 ,7*). The β_{3}/γ_{13} dimer leads to elevated cyctosolic Ca^{2+} levels through phospholipaseC-β_2 generated inositoltrisphosphate (*8-10*). The altered second messenger concentrations of the split pathway are thought, but not yet shown, to generate membrane potential changes and transmitter release. More recently, a family of G protein-coupled receptors (GPCRs), called TAS2Rs, has been identified in mammals (*11, 12*). TAS2Rs colocalize with α-gustducin in TRCs on the tongue with a topography consistent with that of bitter taste receptors. In both, mice and humans, TAS2R genes cluster at chromosomal loci that influence bitter perception (*11*). Functional expression studies and linkage of gene polymorphisms to taste phenotypes clearly identified mouse TAS2R5, as a receptor for the toxin cycloheximide (*13*), an aversive compound in rodents, which is slightly bitter to humans. Binding of cycloheximide to mTAS2R5 activates α-gustducin. Two other receptors, human TAS2R4 and mTAS2R8 responded to denatonium benzoate and 6-n-propyl-2-thiouracil (*13*). However, the extraordinarily high agonist concentrations required to activate these receptors suggest the existence of other, yet unidentified, receptors that mediate the physiological response to these bitter agents (*4*). Together, the above data suggest but do not prove that all TAS2R family members also rcspond to bitter compounds (*3*).

Based on these observations we have isolated all human TAS2R genes and set up a cell based functional expression system to identify activating bitter chemicals for the hTAS2Rs. We show that two of them, hTAS2R16 and hTAS2R10, can specifically be activated by various glycosides and the alkloid strychnine, respectively. Our results give first insights into the interactions of bitter compounds with human bitter receptors and allow us now to develop a screening protocol for bitter masking activities.

Materials and Methods

Molecular biological methods have been described in detail in refs (*14*) (*15*), imaging methods in refs. (*15, 16*), immunocytochemistry in refs. (*15, 17*), and psychophysical methods in refs. (*18-20*).

Identification and Isolation of TAS2R Genes

In order to clone all human TAS2R genes, we searched the human genome data base at the National Center for Biotechnology Information for sequences related to the known mammalian TAS2Rs. Using the TBLASTN algorithm and the known amino acid sequences of mTAS2R5, 8 & 19; rat rTAS2R1-10 & and hTAS2R1, 3-5, 7-10, 13, 14 & 16 we identified 24 human TAS2R genes without introns in their coding regions and 15 pseudogenes at the chromosomal loci 5p15.31, 7p22.3, 7q31.32, 7q34-35 and 12p13.31-13.2 (Figure 1).

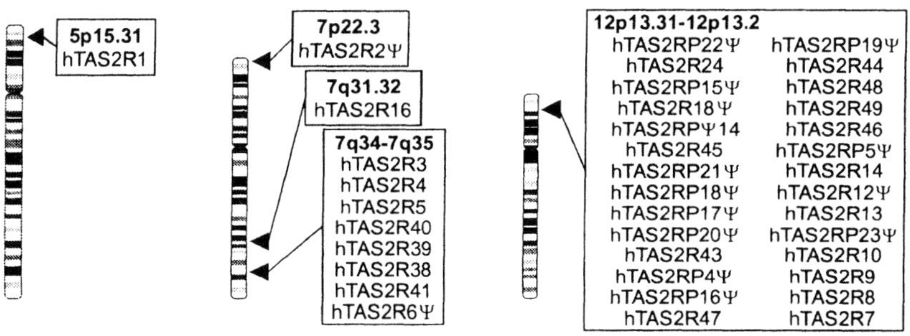

Figure 1: Chromosomal loci of the human TAS2R genes and pseudogenes

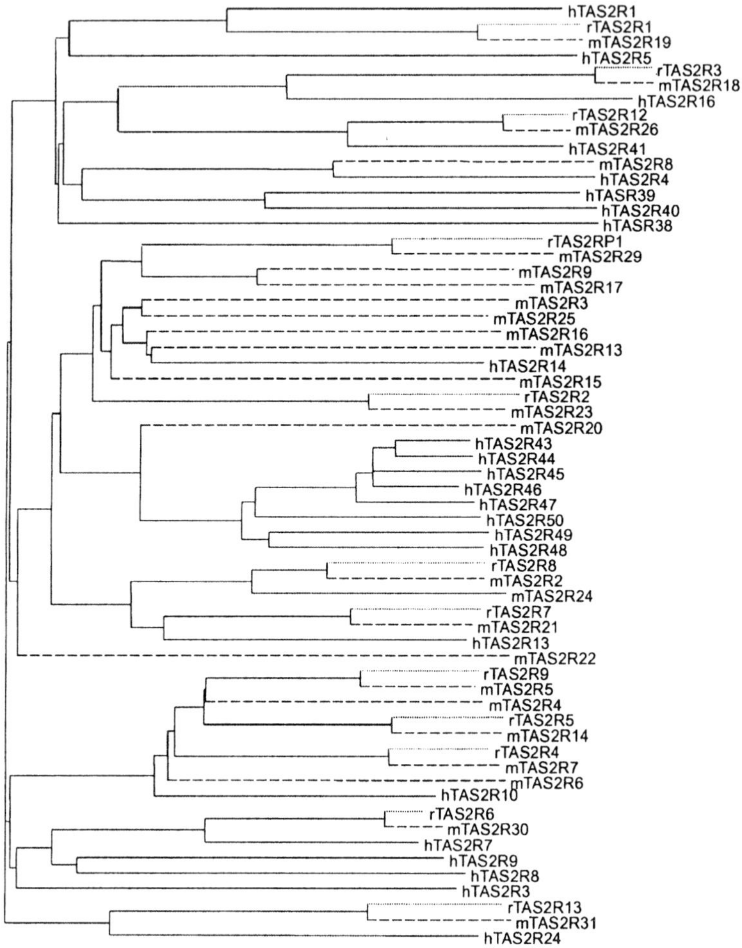

Figure 2: Dendrogram of mammalian TAS2Rs. Solid, dotted and dashed lines depict human, rat and mouse sequences, respectively.

Figure 2 depicts the sequence relationship of the known mammalian TAS2Rs. It is evident that the heterospecic TAS2Rs homologs of the two rodent species are much more similar among one another than to their human counterparts, although the degree of conservation is much lower than that seen for neurotransmitter or peptide hormone receptors. The dendrogram also suggests that the pairs of closely related mouse and rat TAS2R sequences form functional orthologs. However, the pronounced sequence divergence seen between the rodent and human sequences (for example see rTAS2R13/mTAS2R31/hTAS2R24; mTAS2R18/rTAS2R3/hTAS2R16; rTAS2R12/mTAS2R26/hTAS2R41) predicts that the human TAS2Rs do not usually represent functional orthologs of their mice and rat counterparts. Interestingly, the dendrogram also reveals the presence of three human TAS2R subfamilies with the members hTAS2R3, 7, 8 & 9, hTAS2R4 & 38-40, and hTAS2R43-50, and two in the rodent species comprising mTAS2R3, 9, 13, 16, 17, 25 & rTAS2RP1 and mTAS2R4, 5, 7 & 14, rTAS2R 4, 5 & 9. A characteristic of these subfamilies is the occurrence of only one homolog in the other species. The comparatively high sequence divergence between the heterospecic homologous rodent TAS2Rs and the extensive sequence divergence between rodent and human TAS2Rs likely reflect a rapid molecular evolution. This assumption is supported by the appearance of the TAS2R subfamilies, which probably arose by gene amplifications of a common ancestral gene in only one species. We hypothesize that this rapid molecular evolution of the TAS2R genes is associated with the adaptation to different nutritional habits of the species during their phylogenetic evolution.

Construction and analysis of TAS2R expression cassettes

Chemosensory receptors are usually not targeted to the plasma membrane when they are expressed in heterologous cell lines. In order to facilitate plasma membrane targeting, researchers genetically engineered identified targeting signals of other membrane proteins to chemosensory receptors (*13, 16, 21*). We have previously identified a plasma membrane targeting signal in the amino-terminal domain of the rat somatostatin type 3 receptor (rsst3) (*22, 23*). We used this motif to construct a cytomegalovirus promoter-driven TAS2R expression cassette, such that all recombinant TAS2Rs are extended at their amino-termini with the 45 amino-terminal amino acids of rsst3 (figure 3). In order to detect the recombinant receptors with a single monoclonal antibody the expression construct also encoded the herpes simplex virus glycoprotein D epitope at the receptors carboxyl-terminal tails. In previous studies this epitope did not interfere with the expression, trafficking, pharmacological and biochemical properties of some GPCRs (*17, 24*). Next we isolated all 24 human TAS2R coding sequences by PCR experiments from genomic DNA of the human embryonic kidney (HEK)-293 cell line and inserted them into the expression cassette. In similar experiments we obtained several TAS2R constructs from genomic or cDNA sequences of mice and rats.

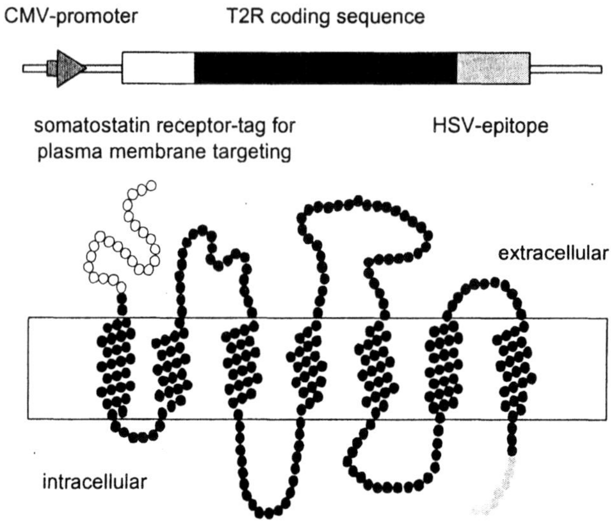

Figure 3: Schematic representation of theTAS2R expression cassette (top) and the chimeric recombinantTAS2R receptors (bottom).

Using liposome-mediated gene transfer we transiently transfected the constructs in HEK293/15 cells and examined the plasma membrane expression of recombinant TAS2Rs. HEK293/15 cells stably express the α-subunit of the mouse G protein G15 (*25*). This G protein has been shown to couple many different GPCRs, including chemosensory receptors, to the inositoltrisphosphate/calcium signaling pathway (*13, 16, 25, 26*). For the analysis of compartmentalization the biotinylated lectin concanavalin A and avidin conjugated to the fluorescence dye Texas Red identified the cell surface glycoproteins. The monoclonal anti-HSV antibody and a secondary anti-mouse IgG antibody conjugated to the fluorescence dye Alexa488 labeled the recombinant TAS2Rs. Cellular images were taken with a confocal microscope. Green fluorescent cells represented the proportion of cells expressing recombinant TAS2Rs and colocalization of both fluorescence dyes indicated cell surface expression of TAS2Rs (Table I). It is apparent that the proportion of cells expressing recombinant receptors varies from 0 to 35 %. Also the extent of cell surface localization differs among the TAS2Rs. In half of the cases the TAS2Rs show clear cell surface localization in all expressing cells while in others cells surface localization is poor or not detectable. Interestingly, cell surface localization of TAS2Rs appears to be unrelated to the expression levels. Although we have not yet examined the biosynthesis of TAS2Rs, low expression levels or absence of expression may indicate high receptor instability and accelerated degradation, while differences in plasma membrane targeting could depend on chaperones with distinct specificity for the various TAS2Rs. More importantly, we note that poor cell surface expression limits the interpretation of data from the functional expression studies (see below).

Table I. Expression of recombinant TAS2Rs in HEK293/15 cells

receptor	% of expressing cells	extent of cell surface localization	receptor	% of expressing cells	extent of cell surface localization
mTAS2R5	2	+	hTAS2R38	35	+
rTAS2R9	27	+	hTAS2R39	14	+/-
hTAS2R1	20	+	hTAS2R40	29	+
hTAS2R3	35	+/-	hTAS2R41	32	+/-
hTAS2R4	not determined		hTAS2R42	0.1	-
hTAS2R5	not determined		hTAS2R43	3	-
hTAS2R7	14	+	hTAS2R44	6	+
hTAS2R8	1	+/-	hTAS2R45	3	-
hTAS2R9	0.5	+/-	hTAS2R46	0.5	-
hTAS2R10	13	+	hTAS2R47	0.0	-
hTAS2R13	16	+/-	hTAS2R48	15	+
hTAS2R14	12	+	hTAS2R49	0.0	-
hTAS2R16	20	+	hTAS2R50	15	+

+, all expressing cells displayed plasma membrane localization of TAS2Rs; +/-, plasma membrane localization of TAS2Rs was detectable, cell surface expression was not seen.

Identification of receptors for salicin and strychnine

To identify bitter compounds that activate TAS2Rs we transfected the constructs encoding all human and several rodent putative bitter taste receptors transiently into HEK293/15 cells grown in quadruplicate in 96-well microtiter plates. 24 h later we loaded the cells with the calcium-sensitive dye Fluo4-acetoxymethylester and recorded calcium responses after bath application of various bitter compounds using the automated fluorescence imaging plate reader (FLIPR) system. Figure 4 illustrates that only a single cell population, i.e. that transfected with hTAS2R16 DNA, showed a significant rise in the cytosolic calcium concentration in response to the β-glucopyranoside salicin.

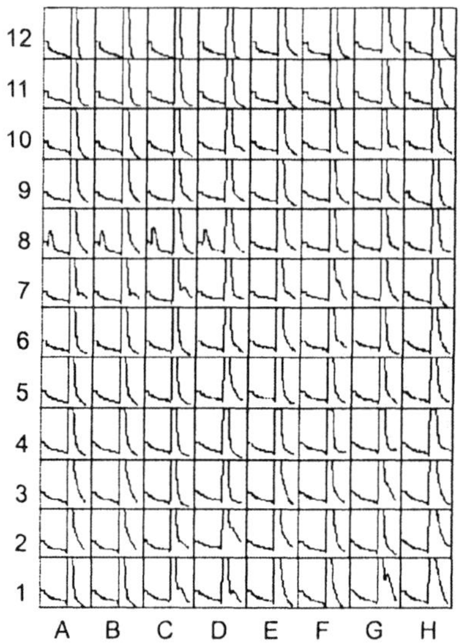

Figure 4. Activation of TAS2R16 by salicin. Cells expressing the 24 human TAS2Rs were grown in quadruplicate on microtiter plates, sequentially stimulated with the bitter compound salicin (1 mM) and isoproterenol (10 μM) and their calcium responses recorded. We used isoproterenol to demonstrate G15 dependent signal transduction. Isoproterenol stimulates an endogenous β-adrenergic receptor that is not normally coupled to the inositoltrisphosphate-calcium signaling pathway. TAS2R16 expressing cells (A8-D8) selectively responded to salicin, while all cells responded to isoproterenol.

All other cells that we previously transfected with the empty cassette or any other TAS2R construct did not respond. In similar experiments we identified another receptor, hTAS2R10, as a responder to the toxic bitter alkaloid strychnine. To verify the FLIPR data we analyzed hTAS2R10 and hTAS2R16 expressing cells by single cell calcium imaging. Figure 5 clearly shows that cells expressing hTAS2R10 responded to strychnine but not salicin, while hTAS2R16 cells reacted to salicin but not strychnine and thereby nicely demonstrates the specificity of the two TAS2Rs. The relatively low proportion of responding cells likely reflects suboptimal signal transduction of the recombinant receptors.

Taken together we have identified two human TAS2Rs as receptors for bitter compounds. We can currently not exclude the possibility of other, yet unidentified receptors for strychnine or salicin. The observed differences in expression levels and plasma membrane localization (Table I) as well as the suboptimal signal transduction of the TAS2Rs in the heterologous expression system (Figure 5) could drop TAS2Rs below the detection threshold. Such receptors would escape identification even if they are exposed to their bitter ligands and currently limit our screening strategy for further bitter receptors.

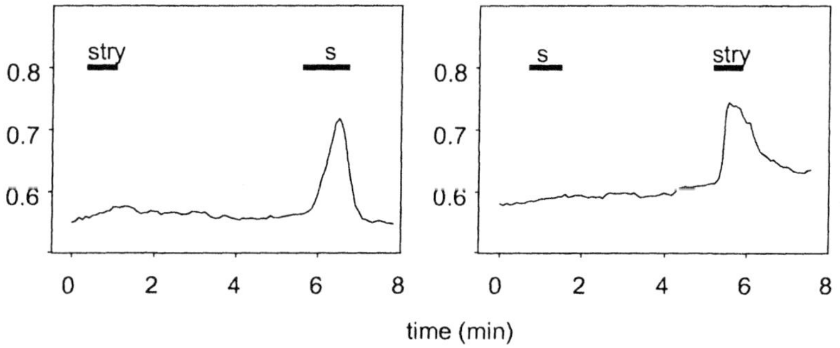

Figure 5: Specificity of the identified bitter receptors. Mean ratio trace of the calcium response in single cells expressing hTAS2R16 (left panel) or hTAS2R10 (right panel) with 5 of 53 and 9 of 81 responding cells. Bars indicate the duration of application of 3 mM salicin and 0.2 mM strychnine.

hTAS2R16 is a broadly tuned receptor for β-pyranosides

Salicin is a β-glucopyranoside with a hydroxymethylbenzene moiety as the aglycon. In order to further characterize the hTAS2R16 we tested several other pyranosides at various concentrations for their ability to activate HEK293/15 cells that express this receptor and calculated the threshold values of activation and the effector concentrations for half maximum responses (EC_{50}), while trained panelists determined in parallel the same parameters for bitter perception of the same compounds (Figures 6 & 7). Notably, the experiments revealed that the parameters for all tested β-glucopyranosides determined in vitro closely resembled those obtained from the psychophysical studies. This accordance provides strong evidence that hTAS2R16 represents a cognate human receptor for bitter β-glucopyranosides. Interestingly, a β-galactopyranoside and an α-glucopyranoside did not activate hTAS2R16, although high concentrations of these substances tasted bitter to humans, suggesting the existence of separate receptors for these substances.

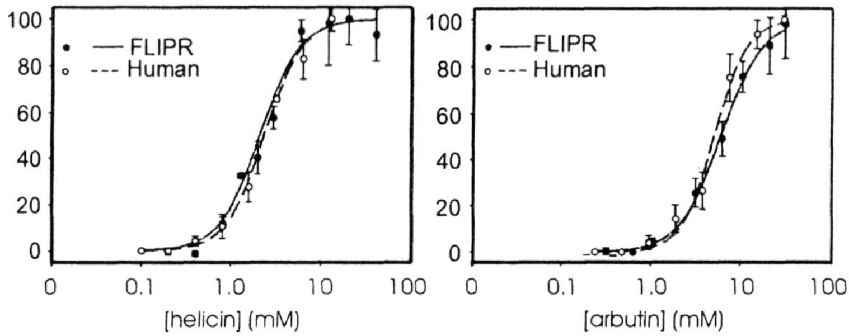

Figure 6. Concentration-response relations of two β-glucopyranosides on the intracellular calcium concentrations in cells expressing hTAS2R16 and on the perception of bitter taste in human subjects

We next analyzed desensitization of hTAS2R16 and adaptation in humans in response to various bitter β-glucopyranosides. Desensitization and adaptation denote diminished responses elicited by chronic or repeated presentation of a stimulus and is frequently observed in sensory systems (*27, 28*). Repeated stimulation of TAS2R16 expressing cells with phenyl-β-D-glucopyranoside resulted in largely diminished responses not only to this compound but also to salicin. This cross-desensitization occurred amongst all tested β-D-glucopyranosides and was fully reversible (Figure 8A). The well-known process of homologous desensitization of agonist-occupied GPCRs, which involves GPCR-specific kinases and arrestins, likely accounts for this result (*28*).

The human subjects made similar observations (Figure 8B). 15 seconds after they took up the test solutions in their mouth, they perceived various β-D-glucopyranoside-TAS2R16 agonists as equally bitter. During prolonged exposure from 30 to 120 seconds to one TAS2R16 agonist they showed adapted bitter perception not only to that initial stimulus, but also to related β-D-glucopyranosides. This cross-adaptation occurred amongst all tested β-D-glucopyranosides, which can activate TAS2R16. Cross-adaptation did not concern the chemically unrelated bitter substance denatonium benzoate, which cannot activate hTAS2R16 (data not shown). The phenyl-α-D-glucopyranoside, which also cannot activate hTAS2R16, failed to cross-adapt with bitter perception of the β-D-glucopyranosides, although its own bitter taste declined strongly (data not shown). We conclude from these results that β-D-glucopyranosides signal bitter taste through a common mechanism involving hTAS2R16, while, in line with our previous assumption, α-D-glucopyranosides activate a separate receptor. Similar observations have recently been made in case of the bitter amino acids L-tryptophan and L-phenylalanine and quinine and urea (*29*). Cross-adaptation occurred amongst the amino acids, but not between the amino acids and quinine or urea indicating that urea and the amino acids activate only partially overlapping bitter taste mechanisms, i.e. they likely activate separate receptors. If all or almost all TAS2Rs are present in the same subset of taste cells (*11*), the above observations suggest that homologous desensitization of TAS2Rs may be involved in bitter taste adaptation.

The data of figure 7 predict some of the structural requirements of compounds to activate hTAS2R16. Both β-configuration of the glycosidic bond and the equatorial orientation of the hydroxyl group at C4 are crucial for receptor activation and perception of bitter taste.
The structure of the aglycons is also important. Large hydrophobic structures mediate better agonist properties than the small methyl group. A hydrogen atom in this position is present in the sweet tasting molecule glucose and prevents agonist activity at hTAS2R16. Hydrophilic substitutions correlate inversely to agonist potency. The glycosidic oxygen atom seems not to be of particular importance because it can be replaced by sulfur. The steric orientation of the hydroxyl group in C2 is of little importance since mannosides act as receptor agonists. Substitutions at C6 had only little effect on agonist properties. The moderate potency of amygdalin, the cyanogenic bitter principle of almonds, is likely due to the hydrophilic nitrile group of its aglycon and not to the α-1,6-glycosidic addition of another glucose molecule. Recent data on the bitter unit of glucopyranosides support our conclusions. This work showed that the aglycons are tasteless and that the bitter unit is formed by a hydrogen acceptor and donor site set up by two of the hydroxyl groups of the glucose moiety (*30*). The bitter unit has been suggested to work like a "hook". It attaches itself to the receptor and pulls the hydrophobic part of the molecule into the receptor site. One of the two is probably the hydroxyl group at C4, since the equatorial position is crucial for TAS2R16 activation. The hydroxyl groups at C2 and C6 likely do not represent the second hydrogen donor/acceptor site, because substations at C6 and the steric position at C2 do not affect activation of TAS2R16. We therefore suggest that the remaining hydoxyl group at C3 represents the second hydrogen acceptor/donor site, although we could not test the commercially unavailable corresponding pyranosides.

Taken together we have identified TAS2R16 as a cognate human bitter receptor for β-glucopyranosides. This receptor links the recognition of a specific chemical structure to the conscious perception of bitter taste. TAS2R16 is tuned to numerous β-glycopyranosides. If this broad tuning is replicated by the other TAS2Rs, it helps explain how humans endowed with only a comparatively small number of bitter receptors can perceive the multitude of bitter compounds. Functional expression of TAS2Rs will be of value for the design and isolation of

compound	structure	threshold values (mM)		EC_{50} (mM)	
		FLIPR	human	FLIPR	human
phenyl-β-D-gluco-pyranoside	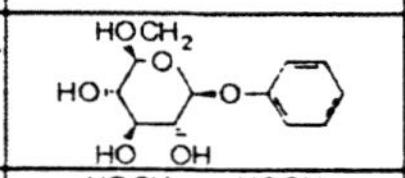	0.07 ± 0.02	0.1 ± 0.05	1.1 ± 0.1	0.7 ± 0.2
salicin	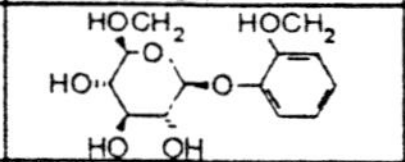	0.07 ± 0.02	0.2 ± 0.1	1.4 ± 0.2	1.1 ± 0.3
helicin	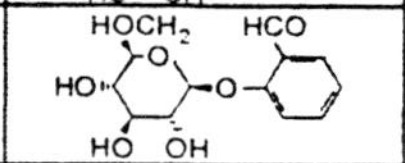	0.3 ± 0.1	0.4 ± 0.1	2.3 ± 0.4	2.2 ± 0.7
arbutin	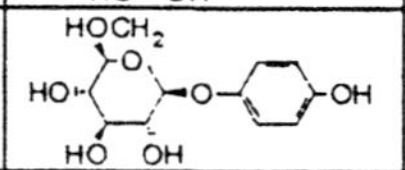	0.5 ± 0.2	0.9 ± 0.3	5.8 ± 0.9	5.4 ± 1.8
2-nitro-phenyl-β-D-gluco-pyranoside	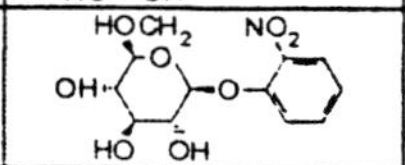	1.5 ± 0.5	n.d.	n.d.	n.d.
naphtyl-β-D-gluco-pyranoside	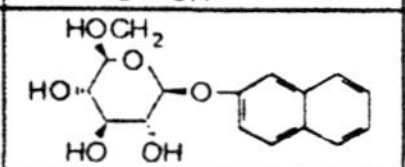	0.4 ± 0.1	0.2 ± 0.1	1.0 ± 0.1	1.4 ± 0.4
amygdalin	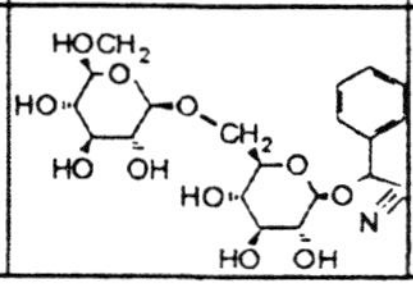	2.3 ± 0.9	n.d.	20 ± 3.4	n.d.
methyl-β-D-gluco-pyranoside	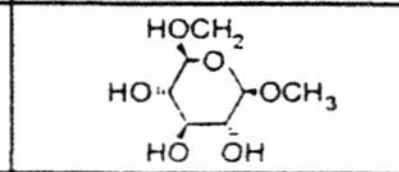	15 ± 6	32 ± 11	n.d.	320 ± 108
esculin	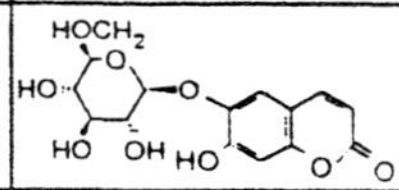	4 ± 2	4 ± 1	n.d.	n.d.
4-nitro-phenyl-β-D-manno-pyranoside	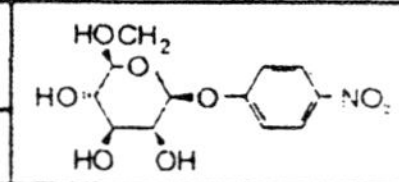	4 ± 2	4 ± 1	n.d.	n.d.
4-nitro-phenyl-β-D-thio-galacto-pyranoside	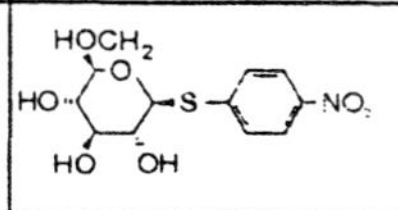	4 ± 2	4 ± 1	n.d.	n.d.
phenyl-β-D-galacto-pyranoside	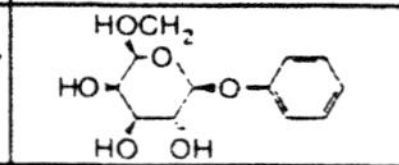	n.r.	40 ± 13	n.r.	n.d.
phenyl-α-D-gluco-pyranoside	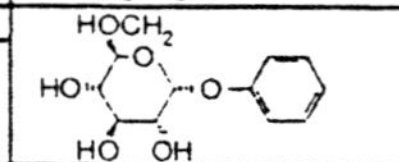	n.r.	9 ± 3	n.r.	50 ± 17
β-D-glucose	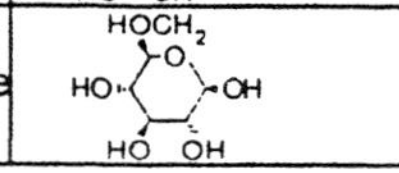	n.r.	sweet	n.r.	sweet

Figure 7. Threshold and EC_{50} values of various pyranosides determined in cells expressing hTAS2R16 and human subjects. n.d., no data due to problems with solubility, osmolarity or toxicity. n.r., no response up to 150 mM.

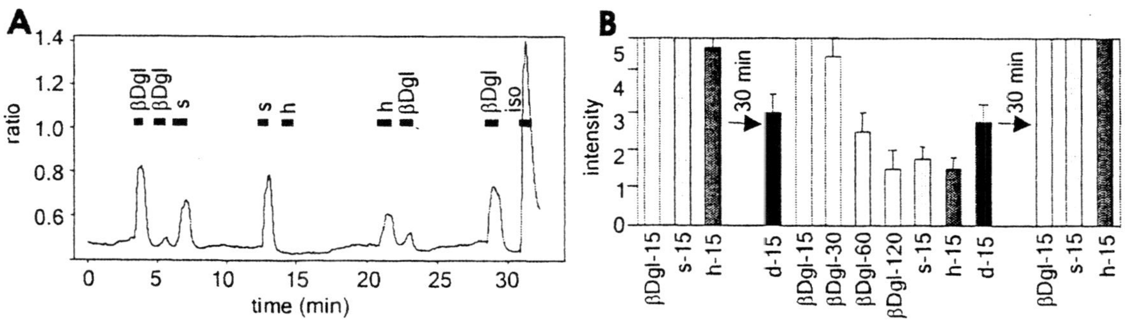

***Figure 8. Cross-desensitization and -adaptation of** β-glucopyranosides-elicited cellular and psychophysical responses.* **A**, *Typical average calcium signals of cells expressing TAS2R16 (6 positive of 60 cells) in response to various pyranosides. Horizontal bars, bath application of 3 mM pyranoside solution. βDgl, phenyl-β-D-glucopyranoside; s, salicin; h, helicin.; 10 μM isoproterenol (iso), positive control.* **B**, *Adaptation of a human test panel to β-glucopyranosides. Bitter intensities of phenyl-β-D-glucopyranoside (βDg, 8 mMl), phenyl-α-D-glucoside (αDgl, 8 mM), salicin (s, 8 mM), helicin (h, 8 mM) and denatonium (d, 0.3 μM) were evaluated at the indicated times (in s).*

bitter masking substances. These could be used to block the bitterness of chemoprotective phytonutrients or medicine and thereby makes them more attractive.

References

1. Drewnowski, A.; Gomez-Carneros, C. *Am. J. Clin. Nutr.* **2000**, 72, 1424-1435.
2. Barratt-Fornell, A.; Drewnowski, A. *Nutr. Today.* **2002**, 37, 144-150.
3. Lindemann, B. *Nature* **2001**, 413, 219-25.
4. Margolskee, R. F. *J. Biol. Chem.* **2002**, 277, 1-4.
5. Wong, G. T.; Gannon, K. S.; Margolskee, R. F. *Nature* **1996**, 381, 796-800.
6. Ruiz-Avila, L.; McLaughlin, S. K.; Wildman, D.; McKinnon, P.J.; Robichon, A.; Spickofsky, N.; Margolskee, R. F. *Nature* **1995**, 376, 80-85.
7. Yan, W.; Sunavala, G.; Rosenzweig, S.; Dasso, M.; Brand, J. G.; Spielman, A. I. *Am. J. Physiol. Cell Physiol.* **2001**, 280, C742-751.
8. Rossler, P.; Kroner, C.; Freitag, J.; Noe, J.; Breer, H. *Eur. J. Cell Biol.* **1998**, 77, 253-261.
9. Huang, L.; Shanker, Y. G.; Dubauskaite, J.; Zheng, J. Z.; Yan, W.; Rosenzweig, S.; Spielman, A.; Max, M.; Margolskee, R. F. *Nat. Neurosci.* **1999**, 2, 1055-1062.
10. Rossler, P.; Boekhoff, I.; Tareilus, E.; Beck, S.; Breer, H.; Freitag, J. *Chem. Senses* **2000**, 25, 413-421.
11. Adler, E.; Hoon, M. A.; Mueller, K. L.; Chandrashekar, J.; Ryba, N. J.; Zuker, C. S. *Cell* **2000**, 100, 693-702.
12. Matsunami, H.; Montmayeur, J. P.; Buck, L. B. *Nature* **2000**, 404, 601-604.
13. Chandrashekar, J.; Mueller, K. L.; Hoon, M. A.; Adler, E.; Feng, L.; Guo, W.; Zuker, C. S.; Ryba, N. J. *Cell* **2000**, 100, 703-711.
14. Stevens, D. R.; Seifert, R.; Bufe, B.; Muller, F.; Kremmer, E.; Gauss, R.; Meyerhof, W.; Kaupp, U. B.; Lindemann, B. *Nature* **2001**, 413, 631-635.
15. Bufe, B.; Hofmann, T.; Krautwurst, D.; Raguse, J.-D.; Meyerhof, W. *Nature Genetics* **2002**, advanced online publication, DOI number 10.1038/Ng1014.
16. Krautwurst, D.; Yau, K. W.; Reed, R. R. *Cell* **1998**, 95, 917-926.
17. Roosterman, D.; Roth, A.; Kreienkamp, H. J.; Richter, D.; Meyerhof, W. *J. Neuroendocrinol.* **1997**, 9, 741-751.
18. Frank, O.; Ottinger, H.; Hofmann, T. *J. Agric. Food Chem.* **2001**, 49, 231-238.
19. Jellinek, G. In *Sensory evaluation of foods - Theory and praxis;* Jellinek, G., Ed; VCH Verlagsgesellschaft: Weinheim, 1985.
20. Mailgaard, M., Civille, G. V., Carr, B. T. In *Sensory evaluation techniques;* Mailgaard, M., Civille, G. V., Carr, B. T., Eds.; CRC Press LLC: New York, N. Y., 1999; pp 122-159.

21. Wetzel, C. H.; Oles, M.; Wellerdieck, C.; Kuczkowiak, M.; Gisselmann, G.; Hatt, H. *J. Neurosci.* **1999**, 19, 7426-7433.
22. Meyerhof, W.; Wulfsen, I.; Schonrock, C.; Fehr, S.; Richter, D. *Proc. Natl. Acad. Sci. U S A* **1992**, 89, 10267-10271.
23. Ammon, C.; Schäfer, J.; Kreuzer, O. J.; Meyerhof, W. *Arch. Physiol. Biochem.* **2002,** 110, 137-145.
24. Kreuzer, O. J.; Krisch, B.; Dery, O.; Bunnett, N. W.; Meyerhof, W. *J. Neuroendocrinol.* **2001,** 13, 279-287.
25. Offermanns, S.; Simon, M. I. *J. Biol. Chem.***1995,** 270, 15175-15180.
26. Bockaert, J.; Pin, J. P. *Embo J.* **1999,** 18, 1723-1729.
27. Torre, V.; Ashmore, J. F.; Lamb, T. D.; Menini, A. *J. Neurosci.* **1995,** 15, 7757-7768.
28. Ferguson, S. S.; Caron, M. G. *Semin. Cell Dev. Biol.* **1998,** 9, 119-127.
29. Keast, R. S.; Breslin, P. A. *Chem. Senses* **2002,** 27, 123-131.
30. Kubo, I. *Physiol. Behav.* **1994,** 56, 1203-1207.

Chapter 4

Genetic Variation in Taste and Preferences for Bitter and Pungent Foods: Implications for Chronic Disease Risk

Beverly J. Tepper[1], Kathleen L. Keller[1,2], and Natalia V. Ullrich[1]

[1]Department of Food Science, Cook College, Rutgers, The State University of New Jersey, New Brunswick, NJ 08901
[2]New York Obesity Research Center, St. Luke's-Roosevelt Hospital Center, Columbia University College of Physicians and Surgeons, New York, NY 10025

Human responses to bitterness and pungency show remarkable diversity, ranging from absolute rejection to strong acceptance. Although the psychobiological processes underlying this variation remain poorly understood, traditional theories point to dietary experience and personality traits as major determining factors. In addition, recent studies on the genetics of taste, specifically the ability to taste the bitter compound 6-n-propylthiouracil (PROP), suggest a major role for this phenotype in the acceptance/rejection of bitter and pungent foods. Many foods contain bitter-tasting and pungent phytochemicals that are known to have antioxidant and chemopreventive properties. Thus, gaining a better understanding of the factors that predict the consumption of such foods is of considerable public health importance.

Bitterness and Human Food Selection

Rejection of bitter taste is ubiquitous in humans and other omnivores including rodents and primates. Since many naturally occurring bitter substances such as plant alkaloids and environmental contaminants are bitter tasting, the avoidance of bitterness may have adaptive significance (*1*). Several lines of evidence support the notion that bitterness is associated with dietary danger. First, humans are exquisitely sensitive to bitter taste and are able to detect bitterness at much lower concentrations than other basic tastes. For example, the detection threshold for quinine solution is in the micromolar range, whereas sucrose and sodium chloride can only be detected at millimolar concentrations (*2*). Second, newborn infants reject bitter tasting solutions suggesting that the physiological mechanisms responsible for detecting and avoiding bitter taste are functional at birth (*3*).

Third, many plant species are capable of synthesizing a variety of bitter and acrid substances that protect them against pathogens and predators (*4-6*). These include such diverse compounds as bitter triterpenes found in citrus fruits; phenolic compounds which are present in citrus, tea, cocoa, grapes and soybeans; and glucosinolates, a large class of organosulfur compounds which are widely distributed in *Brassica* vegetables such as cabbage, turnips, Brussels sprouts and kale. Since many plant-derived compounds also possess anti-nutritional properties or are toxic to humans, bitter taste could provide protection against the consumption of harmful or lethal doses of substances that may be present in human diets (*7, 8*). For example, glucosinolates act as goitrogens by interfering with the uptake of iodine by the thyroid gland and the biosynthesis of thyroid hormone (*8, 9*). Overconsumption of glucosinates is associated with thyroid deficiency and toxic goiter (*10*). A list of selected bitter and pungent phytochemicals and their plant sources appears in Table I.

Interestingly, lower-molecular weight phenolics such as naringin tend to be bitter whereas the higher-molecular weight species (e.g., tannins) are predominantly astringent (*14*). Astringency refers to the sensation of puckering and drying of the surface of the tongue which may be due to precipitation of salivary proteins (*15*). Many phenolics such as the catechins from grape seeds are both bitter and astringent (*16*). Since plant tissues are complex mixtures of heterogeneous substances, both bitterness and astringency may be noticeable in the same material, simultaneously.

Another plant defense mechanism involves the release of volatile isothyiocyanates, commonly called mustard oils (*17*). Upon injury to plant tissues, glucosinolates are hydrolyzed to isothiocyanates. Isothiocyantes have an acrid, cabbage or garlic-like odor to humans and potently stimulate free nerve endings of the mouth, nose and cornea to produce tingling, burning and tearing (*18*). Capsaicin the pungent compound in chili pepper produces similar sensations, but through a different mechanism, namely the activation of vanilloid receptors (*18*).

Table I: Plant Sources of Bitter/Pungent Phytochemicals with Chemopreventive or Antioxidant Properties[1]

Family/Species	Food Source	Representative Compound (sensory quality)
Cruciferae	Broccoli, Cabbage, Brussels Sprouts, Cauliflower	Glucosinolates (bitter)
	Mustard Seed, Horseradish	Isothiocyanates (pungent)
Allium	Onions, Garlic, Scallion, Chives	Diallyl Sulfide, Allyl Methyl Disulfide (pungent)
Vitis	Grapes (juice), Wine	Tannins (astringent)
Citrus	Grapefruit, Lemons, Limes	Naringin (bitter)
Thea	Green Tea, Black Tea	Quercetin (bitter), Epicatechin gallate (bitter, astringent)
Cacao	Chocolate	Catechin, Epicatechin (bitter)
	Soybeans	Genistin (astringent)
Zingiberaceae	Ginger, Tumeric	Gingerol, Curcumin (pungent)
Capsicum	Chili Peppers	Capsaicin (pungent)

[1]Sources: (*11-13*)

Bitter Taste, Pungency and the Consumer

Despite the supposed universality of bitter taste rejection, many commonly-consumed foods and beverages such as fruits, tea, coffee and alcohol have bitterness as a major sensory attribute. Although intense bitterness is generally objectionable, moderate bitterness is a desirable characteristic in these foods. The practice of adding sugar, fat and salt to foods presumably evolved to control bitterness and increase the palatability of these foods (*19*).

Nevertheless, for many consumers, dislike of bitter taste is the primary reason they avoid citrus fruits, bitter green vegetables, and whole-grain products (*11*). Avoidance of these foods is of considerable public health concern since many bitter-tasting and pungent phytochemicals also possess antioxidant and chemopreventive qualities (see Table I). Glucosinolates from cruciferous vegetables have been linked to lower risk of cancer by modulating the activity of Phase 1 and Phase 2 biotransformation enzymes (*20*). Isoflavones from soy and phenolics in green tea may enhance plasma oxidative capacity thereby playing a role in cardiovascular health (*21, 22*).

Dietary policies in the U.S. and elsewhere encourage greater consumption of plant-based foods as a means of reducing the population burden of chronic disease (*24*). The fact that the same compounds that provide chemopreventive benefits are also unpalatable to a large segment of the population seems to place public policy at odds with consumer acceptance behavior (*11*). The food industry routinely employs strategies to reduce bitterness in foods such as selective breeding and commercial de-bittering processes (*11*). However, reduction in bitterness may remove the very same compounds that confer the health benefits. The development of functional foods which minimize objectionable bitterness and preserve or enhance nutritional benefits may be one way to address this problem. Recent advances in the understanding of the receptor mechanisms of bitterness perception could lead to the design of unique compounds that mask or block bitter taste (*24*).

Psychosocial Factors

Individual hedonic responses to bitterness and pungency vary dramatically. Although the reasons for this variation are not well understood, previous research points to dietary experience and personality traits as major mediating variables. A classic example of the influence of dietary experience on food acceptance is the acquired preference for chili pepper. Chili pepper is generally aversive to an individual experiencing it for the first time. However, with repeated exposure, an individual becomes more tolerant of the hotness and may eventually develop a preference for it (*25*). This shift in sensitivity, called desensitization, has been demonstrated in laboratory experiments in which subjects were exposed to capsaicin repeatedly, across days (*26*). These findings are complemented by the results of other studies showing that individuals who consume chili pepper regularly perceive less oral burn from capsaicin treatment than those who rarely consume it (*27*).

There is limited evidence that sensory responses to bitterness are modified by experience. One study found that caffeine users had higher taste detection thresholds for caffeine than non-users (*28*), but another study on alcohol found no effect of alcohol use on sensitivity to the bittering agents in beer (*29*). Moreover, daily exposure to caffeine over a 10-day period in another study was not sufficient to increase acceptance of bitter foods (*19*).

Personality factors may play a unique role in the development of preferences for bitter and pungent foods with physiologic properties, such as alcohol, caffeine and chili peppers. Several studies have shown that those who scored high on a thrill- or sensation-seeking scale showed greater preference for chili pepper and higher consumption of caffeine (*30, 31*) whereas those who scored high on a neophobia (fear of novel foods) scale showed lower preferences for chili pepper and alcohol (*32, 33*).

It is possible that individuals who become chili pepper or coffee consumers are initially less sensitive to the chemosensory effects of these substances and are more likely to accept them. This interpretation suggests that an individual's sensory acuity defines his/her acceptable range of stimulation which may be further modified by personality and experience. As described in the next section, a growing body of work suggests that these perceptual differences may have a genetic basis.

The Genetics of Thiourea Tasting

In 1932, A.L. Fox serendipitously discovered that some individuals are genetically taste blind to a family of bitter tasting compounds with the thiourea moiety (N-C=S). Those designated as thiourea "tasters" perceived moderate to extreme bitterness from these compounds, whereas "non-tasters" found them weak or tasteless (*34*). Two of these compounds, phenythiocarbamide (PTC) and 6-n-propylthiouracil (PROP) have been commonly used in taste research. Their structures are shown in Figure 1. Since PTC was found to have a slight sulfurous odor, PROP has replaced PTC in most contemporary taste studies.

The ability to taste PTC and PROP is a well-studied trait in humans (for review see (*35*)). The frequency of PTC/PROP taste blindness varies considerably worldwide, from ~1% to ~67%. The frequency of the non-taster phenotype among Blacks and Caucasians in the U.S. is estimated at 3% and 30%, respectively (*35*).

It was originally believed that taste blindness to PTC/PROP was controlled by a single genetic locus and was inherited as a simple Mendelian trait. Thus, individuals with both recessive alleles (tt) would be non-tasters and those having one or two dominant alleles would be tasters (Tt or TT) (*36*). Subsequent work suggested that homozygous tasters were more sensitive to PTC than heterozygous tasters (*37*). A substantial body of work supports the notion of two subgroups of tasters, typically referred to as medium tasters and super-tasters (*38*). The trait also shows complex features suggesting that personal and environmental variables may modify the expression of the phenotype. These variables include gender, age, smoking, and the presence of oral disease (*35*). Although the exact mode of inheritance of this phenotype remains a matter of current debate, a number of transmission models have been proposed which

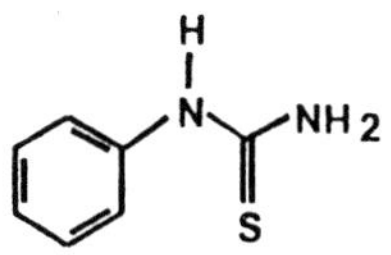

Phenylthiocarbamide (PTC)

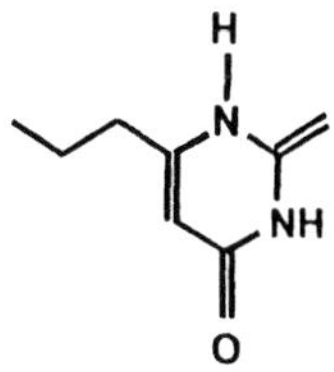

6-n-Propylthiouracil (PROP)

Figure 1. Chemical structure of phenylthiocarbamide (PTC) and 6-n-propylthiouracil (PROP)

include incomplete dominance, variable penetrance and the involvement of multiple alleles or loci (*36*).

The eventual identification and cloning of the receptor(s) will help to resolve this issue. Genetic mapping studies have recently identified a major locus for PTC at human chromosome 5p15 and a second locus on chromosome 7 (*39*). The recent discovery of a cluster of genes expressing bitter receptors in an area showing linkage to PTC (*40*) suggests potential candidates for the PTC receptor or receptors.

Genetic Variation in Responsiveness to Bitterness and Pungency

Studies have shown that PTC/PROP tasters perceive greater intensity from a variety of bitter compounds including solutions of caffeine and urea (*41*), quinine (*42*), potassium chloride, sodium benzoate (*43*), naringin (*44*), and isohumolones (*45*). Delwiche et al. (*46*) recently studied the association between PROP status and perceived bitterness intensity from a panel of 10 bitter compounds including caffeine, quinine, L-amino acids, urea, epicatechin, denatonium benzoate and sucrose octaacetate, among others. Non-tasters gave consistently lower intensity ratings to all compounds studied as compared to medium or super-tasters. However, other studies did not replicate the findings for caffeine, quinine, potassium chloride or sodium benzoate (*41, 42, 47*).

In contrast to the literature on bitter taste, studies examining the association between PROP status and oral irritation have consistently shown that PROP tasters perceive greater irritation from capsaicin, cinnamaldehyde and ethanol (*48-50*). Bartoshuk et al. (*51*) showed that PROP super-tasters were more sensitive to both the bitterness and irritation of ethyl alcohol.

Genetic Variation in Acceptance of Bitter and Pungent Foods

Why the PROP phenotype persists in contemporary human societies is unknown. One classic theory holds that because tasters are more sensitive to bitterness they are more likely to avoid over consumption of dietary goitrogens (*52*). Diets high in goitrogens and limiting in iodine have been associated with increased incidence of goiter and cretinism, an extreme form of thyroid insufficiency leading to mental retardation. The widespread practice of salt iodination has eradicated goiter as a health problem in most developed countries. However, the disorder is still found in isolated populations around the world (*10*). Support for the 'goitrogen-avoidance' hypothesis comes from studies showing that nearly all athyreotic cretins living in an isolated mountain community in the Andes were non-tasters (*52*).

Several studies tested the goitrogen-avoidance hypothesis in healthy populations, but the bulk of this work does not support an association between PROP status and acceptance of cruciferous vegetables or other thiourea-containing foods (*19, 53, 54*). One study found a modest inverse relation between PROP sensitivity and acceptance of cruciferous vegetable in young women (*55*). However, two studies in children reported conflicting results. Keller et al. (*56*) found that PROP taster children gave lower liking ratings to raw broccoli than did non-taster children. However, Anliker et al. (*57*) found no difference in the acceptance of broccoli between taster and non-taster children. Interestingly, other studies have shown that PROP tasters show greater dislike for a range of bitter and strong-tasting foods that do not contain thiourea compounds such as citrus fruits, non-cruciferous vegetables, Japanese green tea, soy products, sauerkraut, strong cheeses, dark beer and black coffee (*55, 58-60*).

Keller (*61*) recently studied the acceptance of citrus juices by children classified by PROP taster status. Fifty-two, 4-5 yr. old children were classified as tasters or non-tasters based on their verbal response to a single solution of PROP (0.56mmol/l) (*56*). Children who stated the solution had no taste were classified as non-tasters (n=18); those who stated the solution tasted bitter, bad or yucky were classified as tasters (n=34). The children tasted samples of orange juice and a 50:50 blend of grapefruit-orange juice in random order. They gave each sample a hedonic rating on a 5-point "smiley-face" scale that ranged from "super bad" to "super good". The experimenter assigned numerical ratings (1=super bad to 5=super good) to the children's responses.

These results are shown in Figure 2. As expected, the liking ratings for grapefruit-orange juice (top panel) were positively skewed for non-tasters and negatively skewed for tasters (χ^2=29.2; p≤0.001). In contrast, liking ratings for the less-bitter, orange juice (bottom panel) were, for the most part, positively skewed for both tasters and non-tasters. These results confirm the findings of an earlier study in young women, which reported lower acceptance of grapefruit juice by PROP tasters (*44*).

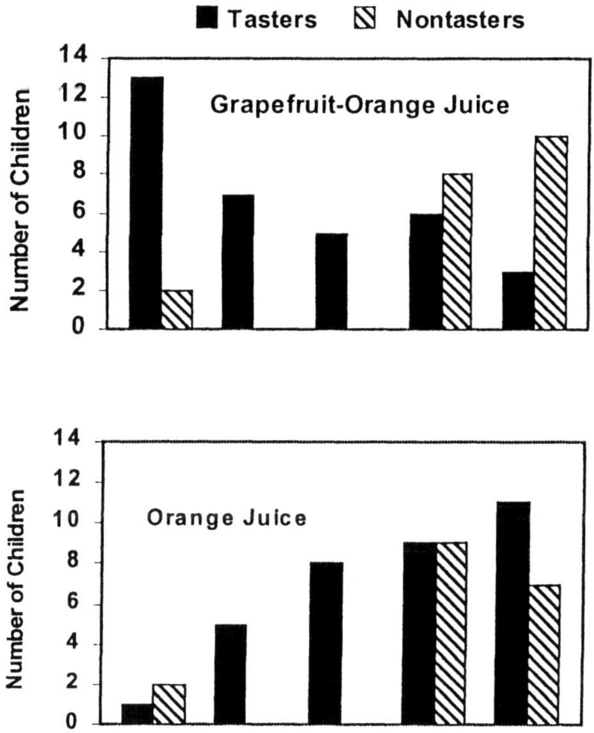

Figure 2. Distribution of liking ratings for grapefruit-orange juice (top panel) and orange juice (bottom panel) by PROP taster (filled bars) and non-taster children (hatched bars).

Role of Food Adventurousness

As stated earlier, personality factors play a unique role in the acceptance of bitter and pungent foods, but they are often ignored in PROP tasting studies. This omission could partially explain why many studies have been unable to relate PROP status to acceptance of bitter and pungent foods.

Ullrich & Tepper (*62*) recently examined the role of 'food adventurousness' in the reported food preferences of PROP tasters and non-tasters. Food adventurousness was defined as the frequency of trying new foods on a scale ranging from 1=never to 4=most of the time. A community sample of 232 adult men and women completed a 65-item food preference checklist indicating whether they liked or disliked each of the foods. Subjects were classified as PROP tasters or non-tasters using a validated method (*63*). Major food groups of interest were vegetables, fruits, alcoholic beverages and pungent condiments. Factor analysis was used to subgroup the foods by preference.

Tasters and non-tasters were then sub-divided according to whether they viewed themselves as food adventurous (willing to try new foods some/most of the time) or not food adventurous (willing to try new foods never/some of the time). This analysis revealed striking differences in the food preferences of PROP tasters as shown in Table II. Tasters who were food adventurous liked more foods in each of the food groups and a greater number of foods overall than did tasters who were not food adventurous. The only exception was the sweet-fruity wine group (rosé wine/wine coolers), which was liked similarly by both groups of subjects. These data revealed that PROP tasters who were also food adventurous liked a wide range of bitter and pungent foods. Only PROP tasters who were not food adventurous displayed the classic dislike of these foods. These differences might have been masked in previous studies that did not distinguish subjects by food adventurousness.

In contrast, non-tasters liked most of the foods on the checklist regardless of whether they were food adventurous or not (data not shown). Thus, non-tasters were little influenced by food adventurousness. Taken together, these findings suggest that genetic taste background in conjunction with food adventurousness provides a more complete picture of food preferences than PROP status alone. Future studies are likely to uncover additional personality traits and behavioral factors that may be linked to food acceptance behavior.

Acceptance of Phytochemicals and PROP Sensitivity – Relationship to Chronic Disease Risk

If PROP tasters avoid many of the foods listed in Table II, they might be at greater risk for low consumption of dietary phytochemicals which could modify their susceptibility to chronic disease. To begin to explore these relationships, Drewnowski & co-workers (*64*) studied the association between PROP status and acceptance of cruciferous vegetables among female breast cancer patients and controls. Results showed a lower acceptance of cruciferous vegetables

Table II: Liking of Vegetables, Fruit, Alcohol and Condiments By Food Adventurous and non-Food Adventurous PROP Tasters [1]

Food Group	*Food Sub-group*	P-value [2]
Vegetables	Turnips, greens, endive, radishes	***
	Cooked carrots, cooked broccoli, raw cabbage	**
	Total	***
Fruit	Grapefruit, grapefruit, cranberry & grape juice, oranges	**
	Strawberries, peaches, bananas	*
	Total	**
Alcohol	Dry red & white wine	***
	Rosé wine, wine coolers	NS
	Gin, bourbon	**
	Dark beer	**
	Total	***
Condiments	Green, black & Greek olives, anchovies	*
	Sour pickles, sauerkraut, raw garlic & onions	**
	Hot peppers, hot sauce, salsa	**
	Total	***

[1] Table II compares the number of foods liked per food sub-group by food adventurous and non-food adventurous PROP tasters. [2] More foods per sub-group were liked by food adventurous PROP tasters at the significance level indicated. * $p<0.05$; ** $p< 0.01$; *** $p< 0.001$.

among all PROP-taster women as compared to all non-taster women. Presumably, if lifelong eating habits alter susceptibility to cancer, then a higher frequency of PROP-tasters might be expected among patients as compared to controls. However, no significant relationships were observed between PROP status and incidence of cancer in this clinical population. Nevertheless, this study represents only an initial first step towards understanding the links between genes, diet and disease and much more works needs to be done before these intricate relationships can be untangled.

A typical strategy employed in epidemiological research is to assess the role of a specific dietary constituent on disease risk. For example, a large number of studies have addressed the specific role of cruciferous vegetables in cancer prevention (*65*), the benefits of wine on cardiovascular health (*66*) and the effects of tea consumption on arterial disease and cancer (*22*). These isolated approaches may not be the most useful for understanding the links between ordinary diets and human health because human diets are diverse, and intake of a specific food is not likely to alter an individual's disease risk.

A final look at Table I reveals that beneficial phytochemicals come from a wide variety of foods including those used as spices such as chili pepper, ginger, tumeric, onions and garlic (*12, 13*). It is possible that spices make a meaningful contribution to total dietary intake of phytochemicals, but their use is rarely assessed in epidemiological studies. A recent study (*5*) surveyed the use of spices in meat-based recipes from 36 countries around the world. Results showed that 65% of recipes called for onion and black/white pepper, 35% called for garlic, and ~25% called for chili pepper or citrus. Confirmation of this intriguing possibility awaits further research.

References

1. Glendinning, J.I., Is the bitter rejection response always adaptive? *Physiol. Behav.* **1994**, 56, 1217-1227.
2. Amerine; M.A., Pang born, R., Rosslea, E.B., *Principles of sensory evaluation of food.* **1965** New York: Academic Press. p. 30-128.
3. Steiner, J.E., Glaser, D. Hawilo, M.E., Berridge, K.C., Comparative expression of hedonic impact; affective reactions to taste by human infants and other primates. *Neurosc.i Behav. Rev.* **2001**, 25, 53-74.
4. Bravo, L., Polyphenols: chemistry. dietary sources, metabolism, and nutritional significance. *Nutr. Rev.* **1998**, 56, 317-333.
5. Billing, J., Sherman, P.W. Antimicrobial function of spices: why some like it hot. *Q. Rev. Biol.* **1998**, 73, 3-49.
6. Fenwick, G.R., Heaney, R.K., Mullin, W.I., Glucosinolates and their breakdown products in food and food plants. *Crit. Rev. Food Sci. Nutr.* **1983**, 18, 123-201.
7. Chung, K.T., Wong, T.Y., Wei, C.I., Huang, Y.W., Lin, Y., Tannins and human health: a review. *Crit. Rev. Food Sci. Nutr.* **1998**, 36, 421-464.

8. Fenwick, G.R., Curl, C.L., Griffiths, N.M., Heaney, R.K., Price, K.R. *Bitter principles in food plants.*, in *Bitterness in foods and beverages; developments in food science*, R.L. Rouseff, Ed. **1990**, Elsevier: Amsterdam. p. 205-250.
9. Cavalieri, R.R., Iodine metabolism and thyroid physiology: current concepts. *Thyroid* **1997**, 7, 177-181.
10. Biassoni, P., Ravera, G., Bertocchi, J., Schenone, F., Bourdoux, P., Influence of dietary habits on thyroid status of a nomadic people, the Bororo sheperds, roaming a central Africa region affected by severe iodine deficiency. *Eur. J. Endocrinol.* **1998**, 138, 681-685.
11. Drewnowski, A., Gomez-Carneros, C. Bitter taste, phytonutrients, and the consumer: a review. *Am. J. Clin. Nutr.* **2000**, 72, 1424-1435.
12. Surh, Y.-J. Anti-tumor promoting potential of selected spice ingredients with antioxidative and anti-inflammatory activities: a short review. *Food Chem. Toxicol.* **2002**, 40, 1091-1097.
13. Munday, R., Munday, C.M. Relative activities of organosulfur compounds derived from onions and garlic in increasing tissue activities of quinone reductase and glutathione transferase in rat tissues. *Nutr. Cancer* **2001**, 40, 205-210.
14. Noble, A.C. Bitterness in wine. *Physiol. Behav.* **1994**, 56, 1251-1255.
15. Horne, J., Hayes, J., Lawless, H.T., Turbidity as a measure of salivary protein reactions with astringent substances. *Chem. Senses* **2002**, 27, 653-659.
16. Robichaud, J.L., Noble, A.C. Astringency and bitterness of selected phenolics in wine. *J. Sci. Food. Agric.* **1990**, 53, 343-53.
17. Govindarajan, V.S. *Pungency: the stimuli and their evaluation.*, in *Food taste chemistry*, J.C. Boudreau, Ed. **1981** American Chemical Society: Washington, DC. p. 53-92.
18. Szallasi, A., Blumberg, P.M. Vanilloid (capsaicin) receptors and mechanisms. *Pharmacol. Rev.* **1999**, 51, 159-211.
19. Mattes, R.D. Influences on acceptance of bitter foods and beverages. *Physiol. Behav.* **1994**, 56, 1229-1236.
20. Talalay, P., Fahey, J.W. Phytochemicals from cruciferous plants protect against cancer by modulating carcinogen metabolism. *J. Nutr.* **2001**, 131, 3027S-3033S.
21. Lichtenstein, A.H. Soy protein, isoflavones and cardiovascular disease risk. *J. Nutr.* **1998**, 128, 1589-1592.
22. McKay, D.L. Blumberg, J.B. The role of tea in human health: an update. *J. Am. Coll. Nutr.* **2002**, 21, 1-13.
23. Potter, J.P. *Food, nutrition and the prevention of cancer: a global perspective.* **1997**, World Cancer Research Fund: Washington, DC.

24. Bufe, B., Hofmann, T., Krautwurst, D., Raguse, J., Meyerhof, W. The human TAS2R16 receptor mediates bitter taste in response to β-glucopyranosides.**2002**, *Nature Genet.* (http://www.nature.com/nature-genetics).
25. Rozin, P., Schiller, D. The nature and acquisition of a preference for Chile pepper by humans. *Motiv. Emotion*, **1980**, 4, 77-101.
26. Green, B.G., Rentmeister-Bryant, H. Temporal characteristics of capsaicin desensitization and stimulus-induced recovery in the oral cavity. *Physiol. Behav.* **1998**, 65, 141-149.
27. Lawless, H.T., Rozin, P., Shenker, J. Effects of oral capsiacin on gustatory, olfactory, and irritant sensations and flavor identification in humans who regularly or rarely consume chili pepper. *Chem. Senses* **1985,** 10, 579-589.
28. Mela, D.J., Mattes, R.D., Tanimura, S., Garcia-Medina, M.R. Relationship between ingestion and gustatory perception of caffeine. *Physiol. Biochem. Behav.* **1992,** 43, 513-521.
29. Guinard, J.X., Zoumas-Morse, C., Dietz, J., Goldberg, S., Holz, M., Heck, E., Amoros, A. Does consumption of beer, alcohol and bitter substances affect bitterness perception? *Physiol. Behav.* **1996,** 59, 625-631.
30. Zuckerman, M., *Sensation seeking: beyond the optimum level of arousal.* 1979, Hillsdale, NJ: Erlbaum.
31. van Trijp, H.C.M., Lahteenmaki, L., Tuorila, H. Variety seeking in the consumption of bread and cheese. *Appetite* **1992,** 18, 155-164.
32. Logue, A.W., Smith, M.E. Predictors of food preferences in adult humans. *Appetite* **1986,** 7, 109-125.
33. Pliner, P., Hobden, K. Development of a scale to measure the trait of food neophobia in humans. *Appetite* **1992,** 19, 105-120.
34. Fox, A.L., The relationship between chemical constitution and taste. *Proc. Natl. Acad. Sci.* **1932,** 18, 115-120.
35. Guo, S.W., Reed, D.R. The genetics of phenylthiocarbamide perception. *Ann. Hum. Biol.* **2001,** 28, 111-142.
36. Snyder, L.H., Inherited taste deficiency. *Science*, **1931,** 74, 151-152.
37. Kalmus, H., Improvements in the classification of the taster genotypes. *Ann. Human Genet.* **1958,** 22, 222-230.
38. Bartoshuk, L.M., V.B. Duffy, I.J. Miller. PTC/PROP tasting: anatomy, psychophysics, and sex effects. *Physiol. Behav.* **1994,** 56, 1165-1171.
39. Reed, D.R., et al., Localization of a gene for bitter-taste perception to human chromosome 5p15. *Am. J. Hum. Genet.* **1999,** 64, 1478-14780.
40. Adler, E., et al., A novel family of mammalian taste receptors. *Cell*, **2000,** 100, 693-702.
41. Hall, M.J., Bartoshuk, L. M., Cain, W. S., Stevens, J.C. PTC taste blindness and the taste of caffeine. *Nature*, **1975,** 253, 442-443.

42. Leach, E.J., Noble, A.C. Comparison of bitterness of caffeine and quinine by a time-intensity procedure. *Chem. Senses*, **1986,** 11, 339-345.
43. Bartoshuk, L.M., Rifkin, B., Marks, L.E., Hooper, J.E. Bitterness of KCL and benzoate: related to genetic status for sensitivity to PTC/PROP. *Chem. Senses*, **1988,** 13, 517-528.
44. Drewnowski, A., Henderson, S. A., Hann, C. S., Shore, A.B. Taste responses to naringin, a flavonoid, and the acceptance of grapefruit juice are related to genetic sensitivity to 6-n- propylthiouracil. *Am. J. Clin. Nutr. 1997,* 66, 391-397.
45. Mela, D.J. Gustatory perception of isohumulones: influence of sex and thiouria taster status. *Chem. Senses*, **1990,** 15, 485-490.
46. Delwiche, J.F., Buletic, Z., Breslin, P.A.S. Covariation in individual's sensitivities to bitter compounds: evidence supporting multiple receptor/transduction mechanisms. *Percep. Psychophysics* **2001**, 63, 761-776.
47. Schifferstein, H.N.J., Frijters, J.E.R. The perception of the taste of KCl, NaCl, and quinineHCl is not related to PROP-sensitivity. *Chem. Senses*, **1991,** 16, 303-317.
48. Tepper, B.J., Nurse, R.J. Fat perception is related to PROP taster status. *Physiol. Behav.* **1997,** 61, 949-954.
49. Karrer, T., Bartoshuk, L.M. Capsaicin desensitization and recovery on the human tongue. *Physiol. Behav.* **1991,** 49, 757-764.
50. Prescott, J., Swain-Campbell, N. Responses to repeated oral irritation by capsaicin, cinnamaldehyde and ethanol in PROP tasters and non-tasters. *Chem. Senses*, **2000,** 25, 239-46.
51. Bartoshuk, L.M. The biological basis of food perception and acceptance. *Food Qual. Pref.* **1993,** 4, 21-32.
52. Shepard, T.H. Phenylthiocarbamide non-tasting among congenital athyroitic cretins: further studies in an attempt to explain the increased incidence. *J. Clin. Invest.* **1961,** 40, 1751-1757.
53. Jerzsa-Latta, M., Krondl, M., Coleman, P. Use and perceived attributes of cruciferous vegetables in terms of genetically-mediated taste sensitivity. *Appetite* **1990,** 15, 127-34.
54. Niewind, A., Krondl, M., Shrott, M. Genetic influences on the selection of Brassica vegetables by elderly individuals. *Nutr. Res.* **1988,** 8, 13-20.
55. Drewnowski, A., Henderson, S. A., Levine, A., Hann, C. Taste and food preferences as predictors of dietary practices in young women. *Public Health Nutr*, **1999,** 2, 513-9.
56. Keller, K.L., Steinmann, L., Nurse, R. J., Tepper, B.J. Genetic taste sensitivity to 6-n-propylthiouracil influences food preference and reported intake in preschool children. *Appetite* **2002,** 38, 3-12.

57. Anliker, J.A., Bartoshuk, L., Ferris, A. M., Hooks, L. D. Children's food preferences and genetic sensitivity to the bitter taste of 6-n-propylthiouracil (PROP). *Am. J. Clin. Nutr.* **1991,** 54, 316-20.
58. Akella, G.D., Henderson, S.A., Drewnowski, A. Sensory acceptance of Japanese green tea and soy products is linked to genetic sensitivity to 6-n-propylthiouracil. *Nutr. Cancer* **1997,** 29, 146-151.
59. Glanville, E.V., Kaplan, A.R. Food preferences and sensitivity of taste for bitter compounds. *Nature* **1965,** 205, 851-852.
60. Intranuovo, L.R., Powers, A.S. The perceived bitterness of beer and 6-n-propylthiouracil (PROP) taste sensitivity. *Ann. NY Acad. Sci.* **1998,** 855, 813-815.
61. Keller, K.L. 2001, Rutgers University: New Brunswick, NJ.
62. Ullrich, N.V., Tepper, B.J. *Food preferences are influenced by gender and genetic taste sensitivity to 6-n-Propylthiouracil. Society for the Study of Ingestive Behavior.* **2000,** Philadelphia, PA.
63. Tepper, B.J., C.M. Christensen, Cao, J. Development of brief methods to classify individuals by PROP taster status. *Physiol. Behav.* **2001,** 73, 571-577.
64. Drewnowski, A., Henderson, S. A., Hann, C. S., Berg, W.A., Ruffin, M. T. Genetic taste markers and preferences for vegetables and fruit of female breast care patients. *J. Am. Dietet. Assoc.* **2000,** 100, 191-197.
65. Murillo, G., Mehta, R.G. Cruciferous vegetables and cancer prevention. *Nutr Cancer* **2001,** 41, 17-28.
66. German, J.B. Walzem, R.L. The health benefits of wine. *Annu. Rev. Nutr.* **2000,** 20, 561-93.

Chapter 5

Genetic Dissection of Sweet Taste in Mice

A. A. Bachmanov, D. R. Reed, X. Li, and G. K. Beauchamp

Monell Chemical Senses Center, 3500 Market Street, Philadelphia, PA 19104

The recent molecular identification and characterization of the T1R taste receptors have dramatically advanced understanding of the sweet taste. Genetic studies of sweetener preferences in mice have substantially contributed to this progress and they remain efficient tools for understanding how sweet taste perception is organized. Several lines of genetic evidence indicate that sweeteners belong to more than one group with potentially different receptor/transduction mechanisms. This is consistent with previous electrophysiological and pharmacological analyses of sweet taste responses and with very recent molecular studies. Analyzing taste responses *in vivo*, combined with *in vitro* molecular approaches, will continue to be a powerful tool for understanding sweet taste perception.

Sugars, and a wide range of other chemicals (referred to here as sweeteners), evoke the sensation of sweetness in humans and are palatable to many other animals. Sweetness perception is initiated in taste receptor cells in taste buds of the oral cavity, and has been presumed to involve the interaction of sweet-tasting compounds with a sweet taste receptor or receptors (*1, 2*).

The consumption of sweeteners requires a complex integration of peripheral sensory, central nervous system and post-ingestive events. Sweeteners not only activate taste receptor cells and gustatory nerves, but they also stimulate release of endorphins (*3-5*) and interact with other brain hedonic mechanisms (*6-10*). In case of nutritive sweeteners, they also provide post-ingestive stimulation (*11*), probably due to their calories. As a complex behavior, consummatory responses to sweeteners are likely to be determined by multiple genes.

Taste responses to sweeteners vary among individual humans, and this variation may be determined genetically (*12-19*). Prominent genetic differences in taste responses to sweeteners also exist among inbred strains of mice. Because mechanisms of taste transduction and genomic sequences are similar in humans and mice, the mouse can be used as a model organism in the studies of the genetics of human taste. The goal of this review is to summarize data available on genetic influences over taste responses to sweeteners in mice.

Mouse Strain Differences in Sweet Taste Responses

Previous studies have shown that inbred mouse strains differ in their avidity for several different sweeteners, including sucrose, glucose, dulcin, saccharin, acesulfame, glycine, D-phenylalanine and L-glutamine (*20-26*). In addition, strain differences in responses of gustatory nerves to sweeteners (*27-29*) and in sweet taste detection assessed using the conditioned taste aversion generalization technique (*30*) were described. These studies indicated that responses to many of these sweeteners (e.g., sucrose, glucose, dulcin, saccharin and acesulfame) closely correlate among mouse strains, but responses to amino acids display somewhat different patterns of strain differences.

In a recent study (*31*), we used mice from 28 genealogically unrelated or distantly related inbred strains and tested them with ascending series of concentrations of four sweeteners using two-bottle tests. The sweeteners were 3 - 30 mM glycine, 3 - 30 mM D-phenylalanine, 1.6 - 20 mM saccharin and 50 - 300 mM sucrose. The responses to sweeteners varied among the mouse strains from indifference to strong preferences. Patterns of strain distributions for different sweeteners were quite distinct, with responses to sucrose and saccharin being closely related, D-phenylalanine showing weaker correlations with these two sweeteners, and responses to glycine being the most dissimilar. This suggests that the strain variation in preferences for these four sweeteners depends on partially overlapping sets of genes. Strain differences in responses to sucrose and saccharin (a nutritive and a non-nutritive sweetener) are most likely determined by identical genetic factors. In contrast, responses to glycine are likely under control of different genetic factors.

Previous studies have shown that mice from the C57BL/6 (B6) strains are among those with the highest preferences for several sweeteners, whereas mice from the 129 strains are among those with the lowest sweetener preferences. Compared with 129 mice, B6 mice had higher preferences for sucrose, saccharin, acesulfame, dulcin, glycine, D-phenylalanine and L-glutamine (*20, 24-26*).

The B6 and 129 mice also differ in preferences for several non-sweet taste stimuli, which allowed us to address whether differences between these two strains are specific to sweet taste, or whether they represent generalized differences in taste responsiveness. Compared with mice from the 129P3/J substrain, mice from the C57BL/6ByJ substrain had higher preferences for citric acid, ethanol, and two umami-tasting compounds, monosodium glutamate (MSG) and inosine 5'-monophosphate (IMP); they had marginally weaker avoidance of quinine, and lower preference scores for NaCl. Preferences for a chemosensory irritant, capsaicin, were similar in the two strains (*32-34*). The lack of generalized differences in taste responses between the two strains is also consistent with previous studies showing that the B6 and 129 strains differed in NaCl consumption (*35-37*), but they both were non-tasters of strychnine, acetates of raffinose, galactose and β-lactose, and they differed only slightly in behavioral responses to cycloheximide and phenylthiourea (*38-41*). This suggests that the differences in sweetener preferences between B6 and 129 mice are not due to generalized taste deficiency in one of the strains. However, these data leave a possibility that the B6 mice have a generalized enhanced avidity for all palatable substances. This possibility was investigated using several approaches as described in the following sections.

The higher consumption of sweeteners, umami-tasting compounds and ethanol by B6 mice, relative to 129 mice, may be due to common underlying mechanisms. One such possible mechanism is caloric value of nutritive sweeteners, umami-tasting compounds and ethanol (e.g., sugars, alcohols and amino acids). To investigate the possibility that higher consumption of these compounds by B6 mice is due to higher caloric or carbohydrate appetite, we measured intake of food, complex carbohydrates (cornstarch and Polycose) and fats (soybean oil and margarine) in these two strains (*42*). In two-bottle tests, compared with 129 mice, B6 mice had higher preferences for dilute Polycose and oil. This was probably due to strain differences in chemosensory perception of these compounds, because at these concentrations, the liquid nutrients had negligible energy content. The higher acceptance of the dilute Polycose solutions by the B6 mice may be explained by greater sweet taste sensitivity of this strain, because Polycose contains approximately 9% glucose and maltose, sugars that have a sucrose-like taste to mice (*29, 30*). The role for sweetness of Polycose is also supported by the fact that the two strains differed in acceptance of Polycose but not starch, a non-sweet carbohydrate with similar nutritive value. No differences between the two strains were found in consumption of mouse chow, margarine, starch, and higher concentrations of Polycose and oil. This suggests that the strain differences in intakes of sweeteners, MSG and ethanol are not caused by differences in appetite for calories. Most likely, the strain differences in preferences for nutritive and non-nutritive sweeteners are due to a common mechanism specific to the sweet taste. Our electrophysiological and genetic studies described below indicate that at least one of these sweet-taste specific mechanisms involves peripheral transduction. The question whether consumption of ethanol and MSG is genetically associated with avidity for sweeteners is addressed in genetic studies described below.

Earlier studies, showing that B6 and 129 mice differ in preferences for non-nutritive sweeteners (*20, 24-26*), indicated that the two strains differ in peripheral processing of sweet taste responses. This was consistent with our demonstration of higher whole nerve chorda tympani responses to sucrose and saccharin for B6 mice compared with 129 mice (*43-45*). To extend these studies, we compared the behavioral and neural responses of B6 and 129 mice to a large number of chemically diverse sweeteners (*27, 46*). The sweeteners tested included sugars (sucrose and maltose), sugar alcohols (sorbitol and erythritol), a chlorinated sugar analog (sucralose), amino acids (D-phenylalanine, D-tryptophan, L-proline and glycine), a dipeptide (aspartame), a protein (thaumatin), N-sulfonyl amides (saccharin and acesulfame-K), a sulfamate (cyclamate), a guanidinacetic acid-based sweetener (SC-45647), a triterpenoid glycoside (glycyrrhizic acid), and a dihydrochalcone glycoside (neohesperidin dihydrochalcone).

Three main patterns of strain differences in sweetener preferences were evident. First, sucrose, maltose, saccharin, acesulfame, sucralose and SC-45647 were preferred by both strains, but the B6 mice had lower preference thresholds and higher solution intakes. Second, the amino acids D-phenylalanine, D-tryptophan, L-proline and glycine were highly preferred by the B6 mice, but not by the 129 mice. Third, glycyrrhizic acid, neohesperidin dihydrochalcone, thaumatin and cyclamate did not evoke strong preferences in either strain. Aspartame was neutral to all 129 mice and some B6 mice, but other B6 mice strongly preferred it. Thus, compared with the 129 mice, the B6 mice had higher preferences for sugars, sweet-tasting amino acids and several but not all non-caloric sweeteners (*46*).

Gustatory afferent responses of the whole chorda tympani nerve were greater in B6 than in 129 mice to the sugars sucrose and maltose, the polyol sorbitol, and the non-caloric sweeteners saccharin, acesulfame, SC-45647, and sucralose. Lower neural response thresholds were also observed in the B6 strain for most of these stimuli. The strains did not differ on their neural responses to amino acids that are thought to taste sweet to mice, with the exception of L-proline, which evoked larger responses in the B6 strain. Aspartame and thaumatin did not evoke neural responses (that exceeded the threshold) in either strain. The strains generally did not differ in their neural responses to NaCl, quinine, and HCl (*27*).

Therefore, both behavioral and neural responses of B6 and 129 mice allowed us to distinguish among three groups of sweeteners: 1) Sugars, sugar alcohols and several artificial sweeteners, which evoked neural responses and preferences in both strains, but for both types of responses, thresholds were lower, and magnitudes were higher in B6 mice; 2) Amino acids, which were preferred by B6 but not 129 mice, and for which neural responses were similar in the two strains (except for L-proline); 3) Aspartame and thaumatin, neither of which were consistently preferred or evoked neural responses in both strains. These data supported the hypothesis that differences in the peripheral gustatory system between the B6 and 129 strains are at least one factor that contributes to differences in their consumption of many sweeteners. Our subsequent studies described below showed that this peripheral gustatory mechanism involves allelic variation of the sweet taste receptor, T1R3.

Although differences in peripheral sweet taste perception appeared to be the major determinant of the strain differences in sweetener preferences, non-sweet sensory (e.g., bitterness, viscosity, osmolality, or coolness resulting from endothermic reactions with saliva) and postingestive factors probably also affected the results for some sweeteners. For example, avoidance of concentrated solutions of saccharin, L-proline and cyclamate may be due to their predominant bitter taste at these concentrations (see also (*47*)). Cyclamate probably does not taste sweet to either mouse strain, as was found in other species (*48, 49*), but the B6 mice may be more sensitive to its bitterness, resulting in a stronger avoidance in this strain compared with 129 mice. The caloric value of the sugars tested (sucrose and maltose) may account for their generally higher consumption compared with the non-nutritive sweeteners (e.g., saccharin, acesulfame, sucralose, SC-45647) in both strains. Sugar alcohols are metabolized differently than sugars, and may also cause discomfort because of their intestinal osmotic effects (*50*). This may have inhibited the mice from consuming large volumes of these solutions and resulted in avoidance of the more concentrated solutions. It is possible that the B6 and 129 mice differ not only in perceived sweetness of the sugar alcohols, but also in their postingestive handling (e.g., intestinal absorption, metabolism, or excretion). Thus, the strain differences in consumption of the sugar alcohols may depend on an interaction between the perception of their sweetness and their postingestive properties.

Genetic Analyses of Sweet Taste Responses in Mice

Analyses of Inheritance

Some early genetic analyses of sweetener consumption by mice yielded evidence that it is influenced by a single locus, named *Sac* (saccharin preference) (*20, 24, 51, 52*), whereas other experiments indicated that more than one gene is involved (*21, 24, 25, 53*). The apparent discrepancy on whether single-gene or multi-gene model better describes genetic variation in sweetener preferences is likely due to use of different progenitor strains and types of mapping panels, different sweetener solutions tested, and different quantitative analyses applied in these studies.

We approached the genetic analyses of sweetener preferences using two parental mouse strains, B6 and 129, in which sweet taste responses have been extensively characterized as described above. In the initial experiment (*32*), we produced F_1 and F_2 hybrids between these two strains and offered them sucrose solution in two-bottle choice tests. Sucrose intakes of the F_1 and F_2 hybrids were significantly higher than the mid-parental value, which demonstrates directional dominance for high sucrose consumption typical of the B6 strain. To assess the number of genetic loci segregating in this cross, we conducted analyses using both quantitative and qualitative trait criteria. The observed F_2 distribution for sucrose intakes differed significantly from the distribution expected from a single-gene model. The single-gene model was also analyzed using qualitative phenotypes of Low (less than 9.2 ml/30 g BW) and High (9.2 ml/30 g BW and greater) sucrose solution intake. The observed frequencies of these qualitative phenotypes in F_2 were significantly different from frequencies expected from the single-gene model. Thus, the single-gene model for sucrose intake was rejected in both the quantitative and qualitative analyses. The trimodal distribution of sucrose intakes in F_2 suggests that this trait is influenced by a few genes with relatively large effects.

Genetic Correlations with Other Taste Stimuli

Because B6 and 129 mice differ not only in sweetener preferences, but also in preferences for ethanol, and sour, salty, umami and bitter compounds, we examined nature of this association in the B6 × 129 F_2 intercross. If the association among these traits in the two parental strains is due to effects of pleiotropic or linked genes, then such traits should also correlate in the segregating F_2 generation. However the association in the parental strains can also be due to fortuitous fixation of several independent genes during inbreeding of these strains. In this case, the traits would segregate among the F_2 individuals independently, and there would be no correlation between the traits.

We found strong positive correlations among preferences and intakes of several sweeteners (sucrose, saccharin and D-phenylalanine), and between sweeteners and ethanol, but no significant correlations between responses to sweeteners and responses to NaCl, quinine or MSG (*32, 34*). This demonstrates that responses to these three sweeteners are largely, but not completely, determined by common genetic factors, and that strain differences in responses to sweeteners are independent from those for salty, bitter or umami stimuli. Although B6 mice avidly consume MSG in amounts comparable with those of sweeteners (*34*), and it has been suggested that MSG may taste 'sweet' to rodents (*54, 55*), these data demonstrate independent genetic determination of strain differences in consumption of sweeteners and MSG.

The genetically determined correlation between sucrose and ethanol intakes is consistent with the hypothesis that the higher ethanol intake by B6 mice depends, in part, on higher hedonic attractiveness of its sweet taste component. The association of sweetener and ethanol consumption and analyses of inheritance for these traits suggest that the differences between B6 and 129 strains for these two traits depend on relatively small and partially overlapping sets of genes (*32*). Results of this and other studies ((*26, 53, 56, 57*), reviewed in (*58, 59*)) support the presence of a genetically determined link between the consumption of ethanol and sweet solutions in rodents. The relationship between ethanol and sweetener perception and consumption is evident from several types of studies. First, electrophysiological recordings indicate that lingual application of ethanol activates sweetener-responsive neural fibers in the gustatory nerves (*60, 61*) and sweetener-responsive units in the nucleus of the tractus solitarius (*62*). Second, conditioned taste aversions generalize between ethanol and sucrose (*63-66*), suggesting that ethanol and sucrose share the same taste property, most likely sweetness. Third, ethanol and sweetener consumption is regulated by common opioidergic, serotonergic and dopaminergic brain neurotransmitter systems (*67-70*). Finally, human alcoholics may have a stronger liking of concentrated sucrose compared with nonalcoholics (*71, 72*).

The *Sac* Locus and Taste Responses to Sweeteners

Positional Cloning of the *Sac* Locus

Using long-term two-bottle tests, Fuller (*51*) has shown that differences in saccharin preferences between the B6 and DBA/2J inbred strains largely depend on a single locus, *Sac*, with a dominant Sac^b allele present in the B6 strain resulting in higher saccharin preference, and a recessive Sac^d allele present in the DBA/2J strain associated with lower saccharin preference. Subsequent studies confirmed this finding in the BXD recombinant inbred strains, in crosses between the B6 and DBA/2 strains (*20, 24, 52, 53, 73*), and in crosses between the B6 and 129 strains (*43*). In addition to sweetener preferences, the *Sac* genotype influences the afferent responses of gustatory nerves to sweeteners (*43-45*) suggesting that the *Sac* gene is involved in peripheral taste transduction and may encode a sweet taste receptor. The *Sac* locus has been mapped to the subtelomeric region of mouse chromosome 4 (*24, 43, 44, 53, 73*).

In a recent study (*74, 75*), using linkage analysis of a B6 × 129 F_2 intercross, and then the marker-assisted selection of a 129.B6-*Sac* congenic strain, we narrowed the region containing *Sac* to a 0.7-cM interval. Then, we constructed a contig of bacterial artificial chromosome (BAC) clones representing the *Sac*-containing region and sequenced a BAC clone including the *Sac* interval. The 0.7-cM *Sac* interval had a physical size of 194 kb and contained twelve predicted genes. Of these twelve genes, only one, *Tas1r3* (taste receptor, type 1, member 3), was a G protein-coupled receptor. Sequence comparison of cDNA from mouse lingual epithelium and genomic DNA showed that *Tas1r3* contains 6 coding exons. It is translated into an 858-amino acid protein with a predicted secondary structure that includes seven transmembrane domains and a large hydrophilic extracellular N-terminus. This structure is typical of the G protein-coupled receptor family 3, which includes the metabotropic glutamate and extracellular calcium-sensing receptors.

A predicted *Tas1r3* protein, T1R3, belongs to a small family of G protein coupled receptors, T1R, which also includes the T1R1 and T1R2 receptors. The three T1R genes (*Tas1r1*, *Tas1r2* and *Tas1r3*) are located on the distal chromosome 4 and are expressed in taste receptor cells (*76-80*). When T1R3 is co-expressed in a heterologous system with T1R2, it acts as a sweet receptor (*81-83*). *Tas1r3* has a human ortholog, *TAS1R3*, residing in a region of conserved synteny in the short arm of human chromosome 1 (1p36) (*84*).

The *Tas1r1* and *Tas1r2* genes were excluded as candidates for *Sac* based on their more proximal chromosomal location (*44, 78, 80*). However the *Tas1r3* gene encoding the T1R3 receptor has been mapped to a more distal part of chromosome 4 corresponding to the *Sac* interval. Based on the effects of the *Sac* genotype on peripheral sweet taste responsiveness (*43-45*), and on involvement of a G protein-coupled mechanism in sweet taste transduction (*2*), *Tas1r3* was selected as the most likely candidate for the *Sac* locus.

If *Tas1r3* is identical to *Sac*, substitution of *Tas1r3* alleles must result in phenotypical changes attributed to the *Sac* locus. Introgression of the 194-kb chromosomal fragment containing the *Tas1r3* allele from the high-sweetener preferring B6 strain onto the genetic background of the 129 strain fully rescued its low sweetener preference phenotype: sweetener intake of the congenic mice was as high as that of mice from the donor B6 strain (*74*). Equivalent 'phenotype rescue' results were obtained in a transgenic experiment (*82*). These data demonstrate that substitution of *Tas1r3* alleles results in behavioral changes attributed to the *Sac* locus and therefore provide strong evidence that *Tas1r3* is identical to *Sac*, and that the T1R3 receptor responds to sweeteners.

Sequence Variants of *Tas1r3*

Identity of *Sac* and *Tas1r3* implies that there must be *Tas1r3* polymorphisms causing variation in sweet taste responses attributed to allelic variants of the *Sac* locus. To identify the *Tas1r3* sequence variants associated with saccharin preference, we analyzed sequences of the *Tas1r3* region in a variety of inbred mouse strains (*85*). We examined genomic sequences including *Tas1r3* exons, introns, upstream and downstream regions, so that polymorphisms affecting amino acid composition, RNA splicing or potential regulatory regions could be detected. To minimize possibility of associations due to common origins, we tested mouse strains with unrelated or distant genealogies. To provide adequate statistical power for detection of phenotype - genotype associations, we analysed 30 mouse strains.

We found a large number of sequence variants among the strains examined. Several of them were significantly associated with saccharin preferences assessed in these strains. These variant sites included protein-coding variants, and variants that did not affect the predicted T1R3 protein sequence (these were a silent mutation in an exon, and those found in the upstream and downstream untranslated regions, and in an intron). All these variants are candidates for functional polymorphisms affecting *Tas1r3* properties. The missense mutations (resulting in non-synonymous amino acid substitution) can affect ligand binding (*86*) or receptor dimerization (*77*). Silent mutations (which do not result in amino acid substitutions) might affect gene or protein function, for instance, by interfering with correct exon splicing (*87*). The polymorphisms in non-coding regions (intronic, and untranslated upstream and downstream) may affect gene expression (*86, 88*). It is possible that the *Sac/Tas1r3* alleles are complex and more than one variant site is needed to affect the phenotype. For example, such complex alleles have been described in fruit flies (*88*). Moreover, it is possible that there are more than two alleles of the *Sac/Tas1r3* gene among the mouse strains examined, so that different polymorphic sites can determine different alleles. The associations found in this study need to be experimentally confirmed in future studies.

Sac/Tas1r3 and Specificity of Taste Reponses to Sweeteners

What are the ligands of T1R3? Two complimentary approaches, *in vitro* studies of heterologously expressed T1R3, and *in vivo* studies of mice with different *Sac* genotypes, shed light on this question.

Recent studies have shown that when expressed in a heterologous system, the mouse T1R2 and T1R3 combination responds to sucrose, fructose, dulcin, saccharin, acesulfame, guanidinacetic acid sweeteners, glycine and several D-amino acids, but not to several sugars (glucose, maltose, lactose, galactose, palatinose) artificial sweeteners (N-methyl saccharin, cyclamate, aspartame and thaumatin) or L-amino acids (*82, 83*). The rat T1R2 and T1R3 combination responded to glucose, fructose, maltose, lactose, galactose, dulcin, saccharin, acesulfame, sucralose, D-tryptophan and glycine, but not to aspartame, cyclamate, monellin, neotame or thaumatin (*81*). The apparently broader range of responses in rats compared with mice can be explained by differences in expression systems between these studies. The human T1R2 and T1R3 combination responded to several sweeteners that did not activate rodent receptors, namely aspartame, cyclamate, monellin, neotame and thaumatin (*81*). This broader responsiveness of the human receptors appears to primarily depend upon the different T1R2, because the combination of human T1R2 and mouse T1R3 responded to aspartame (*83*).

Several aspects of these *in vitro* studies emphasize importance of an *in vivo* approach. First, discrepancies between results obtained using different expression systems leave open a question of whether responsiveness or unresponsiveness to a particular sweetener reflects *in vivo* sensitivity of the receptor, or is an artifact of the *in vitro* system. Second, responses of the heterologously expressed mouse receptors to amino acids were inconsistent with mouse behavioral responses to these stimuli. For example, sweet L-proline and L-threonine did not activate the T1R2 and T1R3 combination, but instead activated the T1R1 and T1R3 combination, which also responded to some umami-tasting and bitter (e.g., L-phenylalanine) compounds (*83*).

Several genetic mapping studies have shown that in addition to saccharin preferences, the *Sac* locus also affects consumption of sucrose, acesulfame and D-phenylalanine (*20, 24, 43, 45, 73*) and also intake of ethanol, which has a sweet taste component (*89*). The *Sac* locus also affects afferent responses of the chorda tympani nerve to sucrose, saccharin and D-phenylalanine (*43-45*). The effects of the *Sac* genotype on taste responses to a variety of compounds suggest that they evoke sweet taste sensation via a common receptor, T1R3.

Our ongoing studies examine specificity of effects of the *Sac/Tas1r3* locus on taste responses to different sweeteners using several approaches. These include analyses of behavioral and neural responses to sweeteners in B6 × 129 F_2 hybrids and in the 129.B6-*Sac* congenic mouse strain, and associations between *Tas1r3* sequence variants and sweetener preferences in multiple inbred mouse strains. The preliminary results indicate that responses to a large number of sweeteners are affected by the *Sac/Tas1r3* allelic variation. However, the effects are sweetener- and concentration-dependent, suggesting that in addition to *Sac/Tas1r3*, other genetic loci play the role in genetic variation of taste responses to sweeteners. These other genes may include additional peripheral taste transduction elements (including novel taste receptors) and/or they may affect central mechanisms of ingestive behavioral responses.

Other Genes Involved In Sweet Taste Responses

Genetic analyses in a number of studies showed multigenic inheritance of sweetener preferences (*21, 24, 25, 32, 53*). Accordingly, several lines of evidence indicated that allelic variation at mouse *Sac/Tas1r3* does not account for all the genetically determined differences in sweetener preferences.

First, the allelic variation at the *Sac/Tas1r3* gene explains only ~3/4 of differences in saccharin preference among 30 inbred mouse strains, and even less so for other sweeteners (*31, 85*). Therefore, substantial genetic variation in sweetener preferences among the strains is attributed to loci other than *Sac/Tas1r3*.

Second, our analyses indicated that the *Sac* locus explains less than 50% of the genetic variance in sucrose intake in the B6 × 129 F_2 cross, showing that more than half of this variation is attributed to other genes (*43*). Consistent with this, in our ongoing mapping studies, we detected several linkages for sweetener preferences at chromosomal locations other than the *Sac* region (unpublished).

One of the genetic loci affecting sweet taste responses is *dpa*, which affects ability of mice to generalize conditioned taste aversion between D-phenylalanine and sucrose, inferring that *dpa* affects ability to detect sweetness of D-phenylalanine. The *dpa* locus also affects responses of sucrose-sensitive fibers of the chorda tympani nerve to D-phenylalanine. B6 mice carry a dominant allele of *dpa* that determines an ability to recognize the sweetness of D-phenylalanine, whereas BALB/c mice carry a recessive *dpa* allele conferring inability to detect D-phenylalanine sweetness. The *dpa* locus was mapped to proximal chromosome 4 (*90, 91*). It was suggested that the *dpa* locus can also affect responses to sweeteners in two-bottle tests (*25*). Consistent with this, we have found a suggestive linkage for sucrose consumption to proximal chromosome 4, in the *dpa* region (*43*). Interestingly, the two loci on the proximal (*dpa*?) and distal (*Sac*) parts of chromosome 4 epistatically interact, suggesting that they may encode interacting components of sweet taste transduction. The non-additive effects of the two loci on sucrose consumption demonstrate that the effect of each locus depends on background genotype, and therefore may vary across mouse strains (*43*).

The other two members of the T1R gene family, *Tas1r1* and *Tas1r2*, also could contribute to genetic variation in responses to sweeteners. We found that the B6 and 129 mouse strains, which differ in sweetener preferences, also differ in *Tas1r1* sequence (*44*). Furthermore, variation between human and mouse orthologs of *Tas1r2* appears to be responsible for the species differences in sweetener responses (*83*). However, there is no evidence that these two genes are associated with differences in sweetener preferences among mouse strains.

Consistent with the mouse work, we have found that strain and individual differences among rats' responses to sweeteners are not related to allelic variation of the rat *Tas1r3* homolog, and therefore must be attributed to the effects of other genes (*92*). Altogether, these data show that sweetener preference has a complex genetic determination.

Conclusion

With the dramatic progress in genetics, chromosomal mapping of traits has become a powerful resource for gene discovery. For taste, genetic mapping of a human PROP sensitivity locus (*93*) and mouse bitter and sweet taste loci (*24, 43, 94*) have resulted in breakthroughs that led to the discoveries of sweet and bitter taste receptors (*74, 77-80, 82, 95, 96*). The data presented in this review demonstrate that taste responses to sweeteners in mice have complex genetic determinations, possibly with responses to different sweeteners affected by different sets of genes. The genes affecting behavioral responses to sweeteners may be involved a number of taste transduction mechanisms, both peripheral and central. Availability of DNA sequence of the mouse genome and other genetic resources make identification of these genes feasible in the near future. This should substantially advance our understanding of mechanisms of sweet taste.

Acknowledgement

Supported by NIH grants R01DC00882 (GKB) and R03DC03853 (AAB).

References

1. DuBois, G. E. In *Firmenich Jubilee Symposium 1895-1995*; Salvadori, G., Ed.; Allured Publishing Corp.: Carol Stream, IL, USA: Geneva, Switzerland, 1995; pp. 32-95.
2. Lindemann, B. *Physiol. Rev.* **1996,** *76,* 719-766.
3. Yamamoto, T.; Sako, N.; Maeda, S. *Physiol. Behav.* **2000,** *69,* 345-350.
4. Marks-Kaufman, R.; Hamm, M. W.; Barbato, G. F. *J. Am. Coll. Nutr.* **1989,** *8,* 9-14.
5. Melchior, J. C.; Rigaud, D.; Colas-Linhart, N.; Petiet, A.; Girard, A.; Apfelbaum, M. *Physiol. Behav.* **1991,** *50,* 941-944.
6. O'Doherty, J. P.; Deichmann, R.; Critchley, H. D.; Dolan, R. J. *Neuron* **2002,** *33,* 815-826.
7. Cooper, S. J.; Barber, D. J. *Pharmacol. Biochem. Behav.* **1994,** *47,* 541-546.
8. Muscat, R.; Kyprianou, T.; Osman, M.; Phillips, G.; Willner, P. *Pharmacol. Biochem. Behav.* **1991,** *40,* 209-213.
9. Schneider, L. H. *Ann. N. Y. Acad. Sci.* **1989,** *575,* 307-319.
10. Yirmiya, R.; Lieblich, I.; Liebeskind, J. C. *Brain Res.* **1988,** *438,* 339-342.
11. Mook, D. G. *J. Comp. Physiol. Psychol.* **1963,** *56,* 645-659.
12. Gent, J. F.; Bartoshuk, L. M. *Chem. Senses* **1983,** *7,* 265-272.
13. Reed, D. R.; Bachmanov, A. A.; Beauchamp, G. K.; Tordoff, M. G.; Price, R. A. *Behav. Genet.* **1997,** *27,* 373-387.
14. Looy, H.; Weingarten, H. P. *Physiol. Behav.* **1992,** *52,* 75-82.
15. Looy, H.; Callaghan, S.; Weingarten, H. P. *Physiol. Behav.* **1992,** *52,* 219-225.
16. Beauchamp, G. K.; Moran, M. *Appetite* **1982,** *3,* 139-152.
17. Desor, J. A.; Greene, L. S.; Maller, O. *Science* **1975,** *190,* 686-687.
18. Greene, L. S.; Desor, J. A.; Maller, O. *J. Comp. Physiol. Psychol.* **1975,** *89,* 279-284.
19. Falciglia, G. A.; Norton, P. A. *J. Am. Dietet. Assoc.* **1994,** *94,* 154-158.
20. Lush, I. E. *Genet. Res.* **1989,** *53,* 95-99.

21. Ramirez, I.; Fuller, J. L. *Physiol. Behav.* **1976,** *16*, 163-168.
22. Pelz, W. E.; Whitney, G.; and Smith, J. C. *Physiol. Behav.* **1973,** *10*, 263-265.
23. Stockton, M. D.; Whitney, G. *J. Comp. Physiol. Psychol.* **1974,** *86*, 62-68.
24. Lush, I. E.; Hornigold, N.; King, P.; Stoye, J. P. *Genet. Res.* **1995,** *66*, 167-174.
25. Capeless, C. G.; Whitney, G. *Chem. Senses* **1995,** *20*, 291-298.
26. Belknap, J. K.; Crabbe, J. C.; Young, E. R. *Psychopharmacol.* **1993,** *112*, 503-510.
27. Inoue, M.; McCaughey, S. A.; Bachmanov, A. A.; Beauchamp, G. K. *Chem. Senses* **2001,** *26*, 915-923.
28. Frank, M. E.; Blizard, D. A. *Physiol. Behav.* **1999,** *67*, 287-297.
29. Ninomiya, Y.; Mizukoshi, T.; Higashi, T.; Katsukawa, H.; Funakoshi, M. *Brain Res.* **1984,** *302*, 305-314.
30. Ninomiya, Y.; Higashi, T.; Katsukawa, H.; Mizukoshi, T.; Funakoshi, M. *Brain Res.* **1984,** *322*, 83-92.
31. Bachmanov, A.; Li, S.; Li, X.; Lu, K.; Tordoff, M.; West, D.; Ohmen, J.; Reed, D.; Beauchamp, G. *Chem. Senses* **2002,** *27*, A95–A96.
32. Bachmanov, A. A.; Reed, D. R.; Tordoff, M. G.; Price, R. A.; Beauchamp, G. K. *Behav. Genet.* **1996,** *26*, 563-573.
33. Bachmanov, A. A.; Tordoff, M. G.; Beauchamp, G. K. *Alcohol. Clin. Exp. Res.* **1996,** *20*, 201-206.
34. Bachmanov, A. A.; Tordoff, M. G.; Beauchamp, G. K. *J. Nutr.* **2000,** *130*, 935S-941S.
35. Bachmanov, A. A.; Tordoff, M. G.; Beauchamp, G. K. *Behav. Genet.* **1998,** *28*, 117-124.
36. Beauchamp, G. K.; Fisher, A. S. *Physiol. Behav.* **1993,** *54*, 179-184.
37. Lush, I. E. In *Genetics of Perception and Communication*; Wysocki, C. J.; Kare, M. R., Eds.; Marcel Dekker: New York, 1991; Vol. 3, pp. 227-241.
38. Lush, I. E. *Chem. Senses* **1982,** *7*, 93-98.
39. Lush, I. E. *Genet. Res.* **1986,** *47*, 117-123.
40. Lush, I. E.; Holland, G. *Genet. Res.* **1988,** *52*, 207-212.
41. Lush, I. E. *Heredity* **1986,** *57*, 319-323.
42. Bachmanov, A. A.; Reed, D. R.; Tordoff, M. G.; Price, R. A.; Beauchamp, G. K. *Physiol. Behav.* **2001,** *72*, 603-613.
43. Bachmanov, A. A.; Reed, D. R.; Ninomiya, Y.; Inoue, M.; Tordoff, M. G.; Price, R. A.; Beauchamp, G. K. *Mamm. Genome* **1997,** *8*, 545-548.
44. Li, X.; Inoue, M.; Reed, D. R.; Huque, T.; Puchalski, R. B.; Tordoff, M. G.; Ninomiya, Y.; Beauchamp, G. K.; Bachmanov, A. A. *Mamm. Genome* **2001,** *12*, 13-16.
45. Bachmanov, A. A.; Reed, D. R.; Ninomiya, Y.; Inoue, M.; Tordoff, M. G.; Price, R. A.; Beauchamp, G. K. *Chem. Senses* **1997,** *22*, 642.
46. Bachmanov, A. A.; Tordoff, M. G.; Beauchamp, G. K. *Chem. Senses* **2001,** *26*, 905-913.
47. Dess, N. K. *Neurosci. Biobehav. Rev.* **1993,** *17*, 359-372.

48. Danilova, V.; Hellekant, G.; Tinti, J. M.; Nofre, C. *J. Neurophysiol.* **1998,** *80,* 2102-2112.
49. Hellekant, G.; Danilova, V.; Ninomiya, Y. *J. Neurophysiol.* **1997,** *77,* 978-993.
50. Schiffman, S. S.; Gatlin, C. A. *Neurosci. Biobeh. Rev.* **1993,** *17,* 313-345.
51. Fuller, J. L. *J. Hered.* **1974,** *65,* 33-36.
52. Belknap, J. K.; Crabbe, J. C.; Plomin, R.; McClearn, G. E.; Sampson, K. E.; O'Toole, L. A.; Gora-Maslak, G. *Behav. Genet.* **1992,** *22,* 81-100.
53. Phillips, T. J.; Crabbe, J. C.; Metten, P.; Belknap, J. K. *Alcohol. Clin. Exp. Res.* **1994,** *18,* 931-941.
54. *Umami: A Basic Taste*; Kawamura, Y.; Kare, M. R., Eds.; Marcel Dekker: New York, NY, 1987; pp. 649.
55. Yamamoto, T.; Matsuo, R.; Fujimoto, Y.; Fukanaga, I.; Miyasaka, A.; Imoto, T. *Physiol. Behav.* **1991,** *49,* 919-925.
56. Overstreet, D. H.; Kampov-Polevoy, A. B.; Rezvani, A. H.; Murelle, L.; Halikas, J. A.; Janowsky, D. S. *Alcohol. Clin. Exp. Res.* **1993,** *17,* 366-369.
57. Stewart, R. B.; Russell, R. N.; Lumeng, L.; Li, T.-K.; Murphy, J. M. *Alcohol. Clin. Exp. Res.* **1994,** *18,* 375-381.
58. Kampov-Polevoy, A. B.; Garbutt, J. C.; Janowsky, D. S. *Alcohol Alcohol.* **1999,** *34,* 386-395.
59. Overstreet, D. H.; Rezvani, A. H.; Parsian, A. *Alcohol Alcohol.* **1999,** *34,* 378-385.
60. Hellekant, G.; Danilova, V.; Roberts, T.; Ninomiya, Y. *Alcohol* **1997,** *14,* 473-484.
61. Sako, N.; Yamamoto, T. *Am. J. Physiol.* **1999,** *276,* R388-396.
62. Di Lorenzo, P. M.; Kiefer, S. W.; Rice, A. G.; Garcia, J. *Alcohol* **1986,** *3,* 55-61.
63. Kiefer, S. W.; Lawrence, G. J. *Chem. Senses* **1988,** *13,* 633-641.
64. Kiefer, S. W.; Mahadevan, R. S. *Chem. Senses* **1993,** *18,* 509-522.
65. Lawrence, G. J.; Kiefer, S. W. *Chem. Senses* **1987,** *12,* 591-599.
66. Blizard, D. A.; McClearn, G. E. *Alcohol. Clin. Exp. Res.* **2000,** *24,* 253-258.
67. Gosnell, B. A.; Majchrzak, M. J. *Pharm. Biochem. Behav.* **1989,** *33,* 805-810.
68. George, S. R.; Roldan, L.; Lui, A.; Naranjo, C. A. *Alcohol. Clin. Exp. Res.* **1991,** *15,* 668-672.
69. Hubell, C. L.; Marglin, S. H.; Spitalnic, S. J.; Abelson, M. L.; Wild, K. D.; Reid, L. D. *Alcohol* **1991,** *8,* 355-367.
70. Pucilowski, O.; Rezvani, A. H.; Janowsky, D. S. *Psychopharmacol.* **1992,** *107,* 447-452.
71. Kampov-Polevoy, A. B.; Garbutt, J. C.; Davis, C. E.; Janowsky, D. S. *Alcohol. Clin. Exp. Res.* **1998,** *22,* 610-614.

72. Kampov-Polevoy, A. B.; Garbutt, J. C.; Janowsky, D. *Am. J. Psychiatry* **1997,** *154*, 269-270.
73. Blizard, D. A.; Kotlus, B.; Frank, M. E. *Chem. Senses* **1999,** *24*, 373-385.
74. Bachmanov, A. A.; Li, X.; Reed, D. R.; Ohmen, J. D.; Li, S.; Chen, Z.; Tordoff, M. G.; de Jong, P. J.; Wu, C.; West, D. B.; Chatterjee, A.; Ross, D. A.; Beauchamp, G. K. *Chem. Senses* **2001,** *26*, 925-933.
75. Li, X.; Bachmanov, A. A.; Li, S.; Chen, Z.; Tordoff, M. G.; Beauchamp, G. K.; de Jong, P. J.; Wu, C.; Chen, L.; West, D. B.; Ross, D. A.; Ohmen, J. D.; Reed, D. R. *Mamm. Genome* **2002,** *13*, 5-19.
76. Hoon, M. A.; Adler, E.; Lindemeier, J.; Battey, J. F.; Ryba, N. J.; Zuker, C. S. *Cell* **1999,** *96*, 541-551.
77. Max, M.; Shanker, Y. G.; Huang, L.; Rong, M.; Liu, Z.; Campagne, F.; Weinstein, H.; Damak, S.; Margolskee, R. F. *Nat. Genet.* **2001,** *28*, 58-63.
78. Montmayeur, J. P.; Liberles, S. D.; Matsunami, H.; Buck, L. B. *Nat. Neurosci.* **2001,** *4*, 492-498.
79. Sainz, E.; Korley, J. N.; Battey, J. F.; Sullivan, S. L. *J. Neurochem.* **2001,** *77*, 896-903.
80. Kitagawa, M.; Kusakabe, Y.; Miura, H.; Ninomiya, Y.; Hino, A. *Biochem. Biophys. Res. Commun.* **2001,** *283*, 236-242.
81. Li, X.; Staszewski, L.; Xu, H.; Durick, K.; Zoller, M.; Adler, E. *Proc. Natl. Acad. Sci. U S A* **2002,** *99*, 4692-4696.
82. Nelson, G.; Hoon, M. A.; Chandrashekar, J.; Zhang, Y.; Ryba, N. J.; Zuker, C. S. *Cell* **2001,** *106*, 381-390.
83. Nelson, G.; Chandrashekar, J.; Hoon, M. A.; Feng, L.; Zhao, G.; Ryba, N. J.; Zuker, C. S. *Nature* **2002,** *416*, 199-202.
84. Reed, D. R.; Li, X.; Chen, Z.; Mascioli, K.; Bachmanov, A. A.; Beauchamp, G. K.; Tordoff, G. K.; Max, M.; Margolskee, R.; Bartoshuk, L. M.; Duffy, V. In *Abstr. of the NAASO meeting*, 2001.
85. Reed, D. R.; Li, S.; Li, X.; Tordoff, M. G.; West, D. B.; Ohmen, J. D.; Beauchamp, G. K.; Bachmanov, A. A. **in preparation**.
86. Rana, B. K.; Shiina, T.; Insel, P. A. *Annu. Rev. Pharmacol. Toxicol.* **2001,** *41*, 593-624.
87. D'Souza, I.; Poorkaj, P.; Hong, M.; Nochlin, D.; Lee, V. M.; Bird, T. D.; Schellenberg, G. D. *Proc. Natl. Acad. Sci. U S A* **1999,** *96*, 5598-5603.
88. Phillips, P. C. *Trends Genet.* **1999,** *15*, 6-8.
89. Bachmanov, A. A.; Reed, D. R.; Li, X.; Li, S.; Beauchamp, G. K.; Tordoff, M. G. *Genome Res.* **2002,** *12*, 1257-1268.
90. Ninomiya, Y.; Sako, N.; Katsukawa, H.; Funakoshi, M. In *Genetics of Perception and Communication*; Wysocki, C. J.; Kare, M. R., Eds.; Marcel Dekker: New York, 1991; Vol. 3, pp. 267-278.
91. Ninomiya, Y.; Higashi, T.; Mizukoshi, T.; Funakoshi, M. *Ann. NY. Acad. Sci.* **1987,** *510*, 527-529.

92. Lu, K.; Tordoff, M. G.; Li, X.; Beauchamp, G. K.; Bachmanov, A. A.; VanderWeele, D. A.; Chapman, C. D.; Dess, N. K.; Reed, D. R. *Chem. Senses* **submitted**.
93. Reed, D. R.; Nanthakumar, E.; North, M.; Bell, C.; Bartoshuk, L. M.; Price, R. A. *Am. J. Hum. Genet.* **1999,** *64,* 1478-1480.
94. Capeless, C. G.; Whitney, G.; Azen, E. A. *Behav. Genet.* **1992,** *22,* 655-663.
95. Adler, E.; Hoon, M. A.; Mueller, K. L.; Chandrashekar, J.; Ryba, N. J. P.; Zuker, C. S. *Cell* **2000,** *100,* 693-702.
96. Matsunami, H.; Montmayeur, J. P.; Buck, L. B. *Nature* **2000,** *404,* 601-604.

Chapter 6

Biomimetic In Vitro Assay for the Characterization of Bitter Tastants and Identification of Bitter Taste Blockers

Stephen A. Gravina[1], Richard A. McGregor[1], Rita Nossoughi[1], John Kherlopian, and Thomas Hofmann[2]

[1]Linguagen Corporation, 2005 Eastpark Boulevard, Cranbury, NJ 08512
[2]Institut für Lebensmittelchemie, Corrensstrasse 45, Westfälische Wilhelms-Universität Münster, D–48149 Münster, Germany

The ability to detect bitter compounds protects against ingestion of noxious and/or poisonous compounds. We are studying guanine nucleotide binding regulatory protein (G protein) activation by bitter compounds utilizing an *in vitro* biochemical assay. This assay is based on the measurement of taste specific G protein activation by isolated taste membranes in the presence of bitter compounds. This assay has been shown to correlate well with the bitterness of a compound and is useful for the identification of bitterness blockers. We have examined multiple compounds isolated from foods that were shown to be bitter by human sensory evaluation. A number of these compounds were shown to activate the gustducin-dependent signal transduction cascade in a concentration dependent manner. Furthermore, we show that the bitter blocker adenosine 5'-monophosphate inhibits the activation of the gustducin-dependent bitter compounds.

Recent advances in the molecular biology and biochemistry of taste have shown that each taste quality affects receptor cells through distinct mechanisms *(1)*. The signal transduction of bitter, umami and sweet compounds is mediated via G protein coupled receptors, while salty and sour tastes depend on ion channels *(2)*.

Gustducin is a taste specific G protein which is homologous to transducin, the G protein of the visual system *(3)*. Gustducin and transducin have been immunocytochemically localized to taste buds and are key elements of bitter and sweet taste signal transduction *(3,4)*. Taste bud containing membranes from bovine circumvallate papillae activated exogenously added gustducin or transducin in response to many bitter stimuli including denatonium benzoate and quinine *(5,6)*. Gustducin has been shown *in vivo* to be implicated in bitter and sweet taste signal transduction *(7)*. Gene replacement was used to generate a null mutation of the α-gustducin gene in mice. The gustducin knockout mice were shown to be deficient in responses to both bitter and sweet tastes as measured by two bottle preference tests as well as electrophysiological responses. The gustducin knockout had no effect on sour or salty responses. Thus, gustducin and transducin-mediated pathways appear to be primary mechanisms by which bitter taste signals are transduced.

Figure 1 shows the current model for bitter taste signal transduction. A bitter molecule interacts with bitter taste receptor's which are localized in the taste bud microvilli. Recently, the bitter taste receptors have been identified as the T2R receptors *(8,9)*. The T2R proteins are a family of of seven transmembrane G protein coupled receptors that have been shown to be activated by bitter compounds and couple to gustducin *(10)*. Thus, activation of T2R receptors by bitter compounds stimulate GTP binding to gustducin and separation of the gustducin α-subunit and gustducin βγ-subunit. The gustducin α–subunit activates phsophodiesterase (PDE) *(4)*. PDE hydrolyzes cAMP which results in the opening of ion channels. The gustducin βγ-subunits activate the phospholipase C cascade which results in mobilization of intracellular stores of calcium *(11)*. The changes in intracellular ions cause release of neurotransmitters, which stimulate neurons to the gustatory central nervous system that recognizes the signal as bitter.

Activation of the bitter signal transduction cascade can be monitored and blocked at the any point in the pathway. We have developed methods to monitor G protein activation by bitter tastants. G protein activation can be evaluated by binding of the nonhydrolysable GTP analogue [^{35}S] GTPγS. Since gustducin and transducin have been shown to be biochemically interchangeable, we utilize transducin as the reporter *(4, 5)*. We have used transducin to develop useful *in vitro* biochemical assays that correlate with *in vivo* perceived bitterness and bitterness blockers. We have adapted this technology to a high throughput screening format to identify novel bitterness blockers and to identify bitter compounds isolated from complex mixtures.

Herein we report the validation of the biomemitic *in vitro* bitter assay. Furthermore we utilize this method to analyze the bitterness of isolated Maillard compounds and other bitter pharmaceuticals. Result show time and

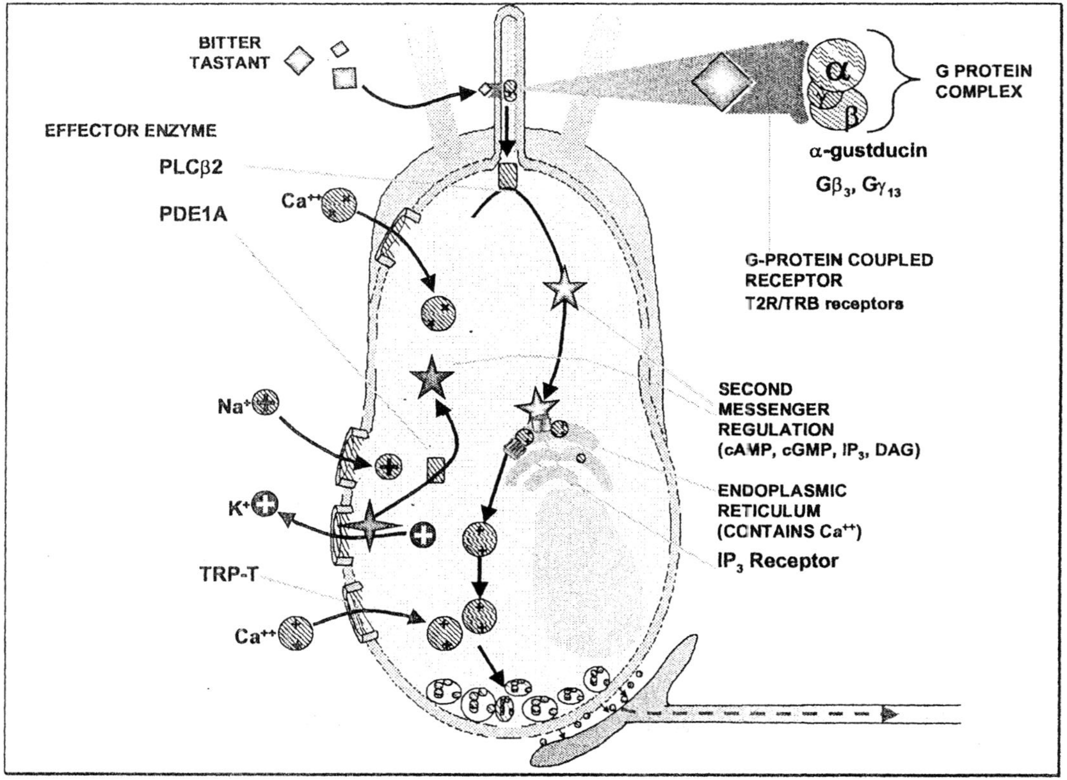

Figure 1. Bitter taste signal transduction pathways.

concentration dependent activation of G proteins in response to bitter compounds in the presence of circumvallate membranes.

Materials and Methods

Chemicals

Dextromethorphan HBr was obtained from the pharmaceutical industry. All other bitter compounds were either obtained from Aldrich (Steinheim, Germany), or were isolated from foods (Hofmann, unpublished results), and were purified by recrystallization or chromatography. Prior to sensory analysis, the purity of the synthetic taste compounds was proven by LC/MS and ^{1}H NMR spectroscopy. All taste compounds were dissolved in tap water and adjusted to pH 7.0 with diluted hydrochloric acid or NaOH.

Radiometric [^{35}S] GTPγS Binding Assay.

Following a procedure reported earlier, the binding of [^{35}S] GTPγS to transducin in bovine circumvallate membranes was measured in the absence (control) and presence of bitter tastants, respectively *(12)*. After incubation, samples were filtered through nitrocellulose filters, washed with buffer, the filters were dried, placed in scintillation fluid and the radioactivity was determined by scintillation counting. Experiments were performed in triplicate.

Mouse Two-Bottle Preference Assay

The two-bottle preference test was performed as follows *(10)*. Each test group contained 10 mice (5 females, 5 males), housed individually and fed *ad libitum.* Mice were presented with distilled water in two sipper bottles 48 hours before the tests, to avoid novelty artifacts. During the 48-hour test period, mice were provided with two bottles with various test solutions. After 24 hours the bottle positions were switched to control for positional cues. After each 24-hour period, the volumes consumed were recorded and the bottle positions were switched.

Sensory Panel Training

Assessors were trained to evaluate the taste quality of aqueous solutions (3 mL each) of the following standard taste compounds by using a triangle test:

saccharose (50 mmol/L) for sweet taste; lactic acid (20 mmol/L) for sour taste; NaCl (12 mmol/L) for salty taste; caffeine (1 mmol/L) for bitter taste; monosodium glutamate (MSG, 8 mmol/L, pH 5.7) for umami taste; tannin (gallustannic acid; 0.05%) for astringency. In addition, the panelists were asked to evaluate the taste intensity and the time/response function of aqueous solutions containing increasing amounts of these taste compounds. Sensory analyses were performed in a sensory panel room at 22-25°C on three different sessions.

Results

Membranes derived from bovine circumvallate papillae were incubated with transducin, [^{35}S] GTPγS and various bitter compounds. Bitter compounds stimulate bitter-responsive GPCRs in the membrane, which activates [^{35}S] GTPγS binding to transducin. Free [^{35}S] GTPγS was separated from transducin-bound [^{35}S] GTPγS by filtration and the total counts per minute (CPM) were measured by scintillation counting. Figure 2 shows the time course of

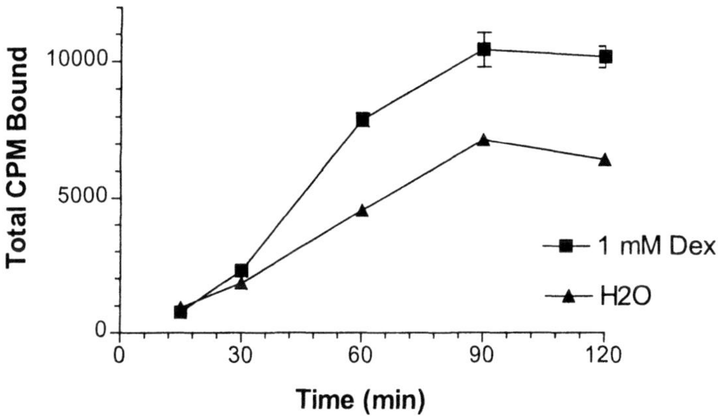

Figure 2. Time course of G protein activation by circumvallate membranes in the presence or absence of dextromethorphan HBr. The reaction was terminated at 15, 30, 60, 90 or 120 minutes by filtration and the filter-bound radioactivity was measured by scintillation counting. Each data point represents the mean of three separate experiments ± SEM.

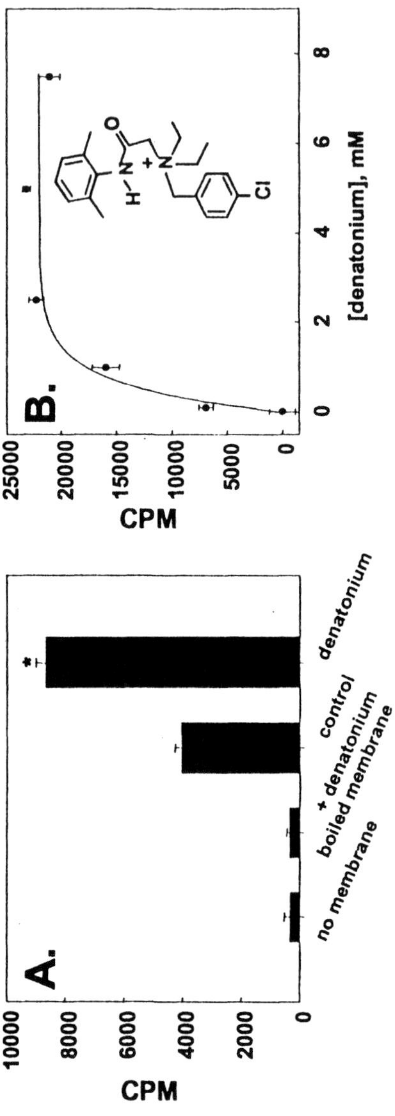
A.
CPM
10000
8000
6000
4000
2000
0
no membrane
boiled membrane + denatonium
control
denatonium
*
B.
CPM
25000
20000
15000
10000
5000
0
0
2
4
6
8
[denatonium], mM
Cl

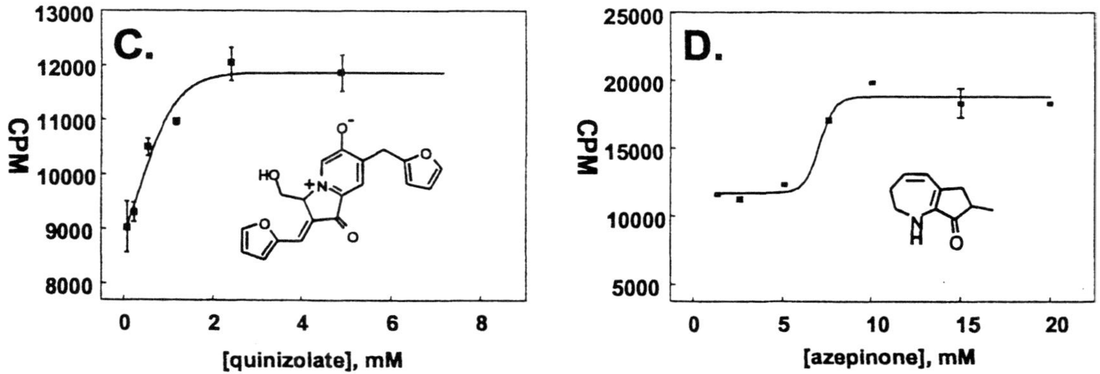

Figure 3. A. [^{35}S] GTPγS binding in the presence or absence of denatonium benzoate (1mM). Bars indicate the effects of no circumvallate membranes + denatonium, boiled membranes + denatonium, no denatonium and + denatonium respectively. B, C, D. Concentration dependence of denatonium benzoate, quinizolate and azepinone. Each data point represents the mean of three separate experiments ± SEM.

transducin activation in the presence (closed squares) or absence (closed triangles) of the bitter cough suppressant dextromethorphan HBr (1 mM). The time course data show that in the absence of bitter compound there is a significant background of [^{35}S] GTPγS binding to transducin; however, the rate of binding is increased in the presence of dextromethorphan. The apparent long incubation times required to measure specific signals are similar to what is seen with diluted solutions of light activated rhodopsin, the natural GPCR for transducin. Thus bitter-responsive GPCRs are at relatively low concentrations in our isolated membrane preparation. Transducin is known for it's extremely low non-specific rate of activation in the presence of dark-adapted rhodopsin (data not shown). The high background observed in the time course may be a property of bitter-responsive GPCRs or a response to one of the compounds present in the assay buffer.

Figure 3 A shows control experiments for [^{35}S] GTPγS binding in the presence or absence of the bitter compound denatonium benzoate. Bar one (no membrane) shows the results of [^{35}S] GTPγS binding in the absence of membranes, but in the presence of denatonium. In contrast to the control no bitter compound results shown above, little [^{35}S] GTPγS binding is observed in the absence of circumvallate membranes. Thus, activation of endogenous gustducin cannot be measured by this method. Bar two (denatonium + boiled membranes) shows that transducin activation is dependent upon intact membranes.

Bar three (control) shows similar non-specific activation in the absence of denatonium that is increased in the presence of denatonium (Bar 4). Figure 2 B shows denatonium concentration dependence of [^{35}S] GTPγS binding. Non-specific binding was subtracted from specific to give Δ CPM values. These data show typical saturation curve kinetics for denatonium with an EC50 of ~0.5 mM. The EC50 value is in good agreement with that observed with denatonium (1 mM) stimulation of bovine chorda tympani nerves *(13)*. It is significantly higher that that observed for human taste threshold (~10 nM). This discrepancy may be due to species differences in bitter taste receptors. Figure 3 C and D show the concentration dependence of quinzolate and azepinone. These compounds were isolated from foods and are Maillard reaction products. The EC50 values for quinizolate and azepinone were 1 and 7.5 mM respectively.

Figure 4 shows [^{35}S] GTPγS binding data for eight Maillard reaction products. Values represent the maximal binding data obtained from concentration dependence studies. Each compound maximally stimulated [^{35}S] GTPγS binding to differing degrees. The different maximal binding between compounds may be due to stimulation of different subtypes of bitter receptors.

Previously adenosine 5'-monophosphate (AMP) was discovered to be a blocker of the gustducin/transducin-dependent bitter taste cascade using a trypsin-digest assay similar to the [^{35}S] GTPγS binding assay *(14)*. We examined AMP's effect *in vivo* utilizing a mouse two-bottle preference test with dextromethorphan (Figure 5).

Bar 1 shows the preference ratio for water. As expected the preference ratio is nearly 0.5 suggesting that there was no preference for either bottle of water. Bar 2 shows the results with dextromethorphan (1 mM) vs. water. The negative correlation suggests that the mice avoided the bitter dextromethorphan bottle. Bar 3 shows the results with dextromethorphan + AMP *vs.* water. The positive preference ratio for the dextromethorphan + AMP demonstrate that this bottle was preferred *vs.* water and suggest that AMP has blocked the bitter taste of dextromethorphan.

To examine the effects of AMP on human bitter taste perception, caffeine solutions were evaluated with increasing amounts of AMP (Figure 6). These data show that increasing quantities of AMP (0-1 mM) will sequentially decrease the perceived bitterness of caffeine solutions. The final bar shows that after the appropriate rest period the perception of bitter returns to normal.

Conclusions

We have developed a high throughput method based on the reconstitution of the gustducin/transducin-dependent bitter taste signal transduction cascade. This method was useful in measuring the time and concentration dependence of a variety of bitter compounds. Moreover, this method can be utilized to identify bitter blockers. A first generation version of the GTPγS binding assay led to the

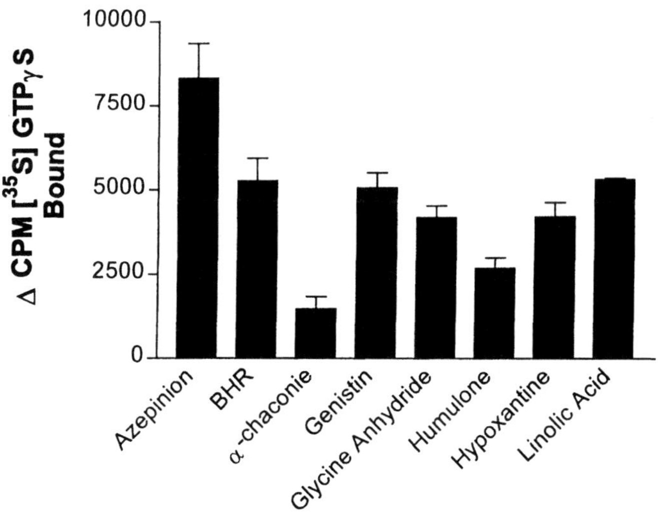

Figure 4. Stimulation of [^{35}S] GTPγS binding to G protein reporter by bitter compounds isolated from foods. Results shown were obtained from the maximal [^{35}S] GTPγS binding derived with concentration dependence experiments.

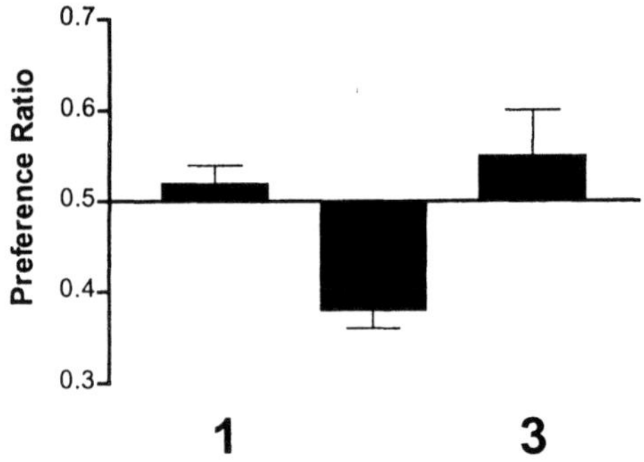

Figure 5. Mouse Two Bottle Preference Test. Bar 1 Water, bar 2 dextromethorphan HBR (1 mM), bar 3 dextromethorphan HBr (1 mM) + adenosine 5'-monophosphate (1 mM). The volume of liquid in each bottle was measured at 24 and 48-hours. The preference ratio is calculated as = bottle B ml/(bottle A ml + bottle B ml).

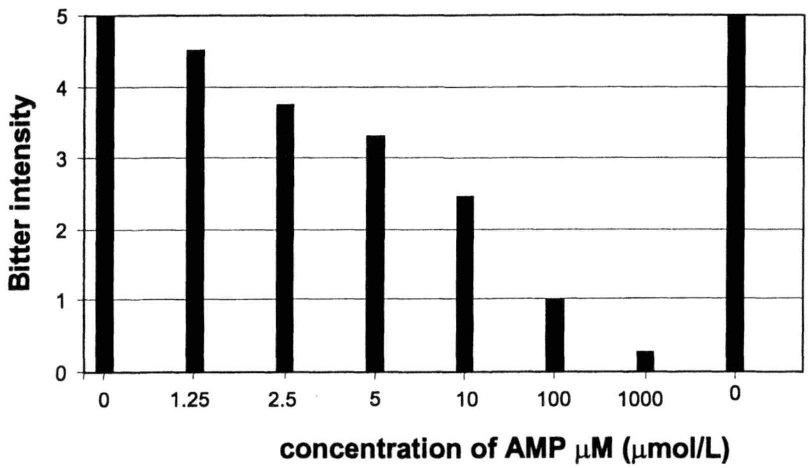

Figure 6. Human sensory evaluation of AMP with caffeine. Subjects were presented with solutions of caffeine (2.5 mM), with various concentrations of AMP (0-1 mM). Subjects utilized a sip and spit method to taste the mixtures and were asked to rate on a scale of 1-5 the bitter intensity.

identification of AMP as a bitter blocker *(6)*. We have extended this work to show that AMP blocks the bitter taste of dextromethorphan and caffeine *in vivo* in mice and humans respectively.

References

1. Smith, D.S.; Margolskee, R.F. *Scientific American* **2001,** 284, 1-9.
2. Gilbertson, T.A.; Damak, S.; Margolskee, R.F. *Current Opinions in Neurobiology* **2000**, 4, 519-27.
3. McLaughlin, S.K.; McKinnon, P.J.; Margolskee, R.F. *Nature* **1992,** 357, 563-569.
4. Ruiz-Avila, L.; McLaughlin, S.K.; Wildman, D.; McKinnon, P.J.; Robichon, A.; Spickofsky, N.; Margolskee, R.F. *Nature* **1995,** 376, 80-84.
5. Hoon, M.A.; Northup, J.K.; Margolskee, R.F.; Ryba, N.J.P. *Biochem. J.* **1995,** 309, 629-636.
6. Ming, D.; Ruiz-Avila, L.; Margolskee, R.F. *Proc. Natl. Acad. Sci. USA* **1998,** 95, 8933-8938.
7. Wong, G.T.; Gannon, K.S.; Margolskee, R.F. *Nature* **1996,** 381, 796-800.
8. Adler, E.; Hoon, M.A.; Mueller, K.L.; Chandrashekar, J.; Ryba, N.J.P.; Zuker, C.S. *Cell* **2000,** 100, 693-702.
9. Matsunami, H.; Montmayeur, J.P.; Buck, L.B. *Nature* **2000,** 404, 601-604.
10. Chandrashekar, J.; Mueller, K.L.; Hoon, M.A.; Adler, E.; Feng, L.; Guo, W.; Zuker, C.S.; Ryba, N.J.P. *Cell* **2000,** 100, 703-711.
11. Huang, L.; Shanker, Y.G.; Dubauskaite, J.; Zheng, J.Z.; Yan, W.; Rozenzweig, S.; Spielman, A.I.; Max, M.; Margolskee, R.F. *Nat. Neurosci.* **1999,** 12, 1055-1062.
12. Asano, T.; Pedersen, S.E.; Scott, C.W.; Ross, E.M. *Biochemistry* **1984,** 23, 5460-5467.
13. Danilova, V.; Hellekant G. 23rd Annual Meeting of the Association for Chemoreception Sciences. April 25-29, 2001.
14. Ming, D.; Ninomiya, Y.; Margolskee, R.F. *Proc. Nalt. Acad. Sci. USA* **1999,** 96, 9903-9908.

Analytical Characterization and Structure–Activity Relationships

Chapter 7

The Taste Activity Concept: A Powerful Tool to Trace the Key Tastants in Foods

Thomas Hofmann[1], Harald Ottinger[2], and Oliver Frank[2]

[1]Institut für Lebensmittelchemie, Corrensstrasse 45, Westfälische Wilhelms-Universität Münster, D–48149 Münster, Germany
[2]Deutsche Forschungsanstalt für Lebensmittelchemie, Lichtenberstrasse 4, D–85748 Garching, Germany

In order to get more insight into the structures and sensory activities of those tastants which are not present in the foods *per se*, but are generated during food processing by Maillard-type reactions from carbohydrates and amino acids, remain unknown. In order to rank the tastants according to their relative taste impact and to identify the key tastants generated during thermal food processing, the so-called Taste Dilution Analysis (TDA) using the human tongue as a biosensor for tastants, was applied on heated, intensely bitter tasting pentose/amino acid solutions. This screening technique led to the identification of previously unknown quinizolate and homoquinizolate exhibiting extraordinarily low bitter thresholds of 0.25 and 1.0 μmol/kg water, respectively. Finally, their taste contribution was estimated by calculating their taste activity values as the ratio of their concentrations to their human taste detection thresholds. This so-called taste activity concept revealed that 73.0 % of the overall bitterness of the reaction mixture was accounted for these two tastants only.

The sensory impression "flavor" is due to the simultaneous stimulation of the human chemical senses odor and taste, and is triggered by chemical compounds naturally present in food products. Although the consumer acceptance of foods is strongly influenced by aroma-active volatiles as well as by non-volatile, taste-active compounds, the flavor research in the last decades was focused mainly on the volatiles. The development of screening procedures using the human nose as an odor-sensitive biosensor, such as, e.g. gas chromatography/olfactometry (*1*), Charm[R] analysis (*2*), or aroma extract dilution analysis (*3*) has enforced the straightforward identification of the key odorants in various foods. Furthermore, the determination of the so-called odor activity values as the ratio of the "natural" concentration and the odor detection threshold of the odorants identified opened the possibility to estimate the contribution of each odorant to the overall aroma of foods such as, e.g. rye bread crust (4), roasted sesame (5), strawberry juice (6), or orange juice (7).

Compared to aroma-active volatiles, relatively little attention has been paid to tongue responses induced by non volatiles. In analogy to the aroma activity value, the taste activity value (TAV) was recently defined as the ratio between the concentration and the taste threshold concentration of a compound in order to evaluate the contribution of individual taste compounds to the overall taste of foods (*8-9*). The evaluation of the sensory importance of individual taste compounds in foods on the basis of their TAVs was, however, as yet mainly focused on compounds, which were already known in structure. E.g. the TAVs of mineral salts, amino acids, nucleotides, and peptides etc. were determined in order to estimate their contribution to swiss cheese (*8*) or stewed beef juice (*9*).

In particular, the structures and sensory activities of intense tastants which are not present in the foods *per se*, but are generated during food processing from precursors, e.g. by Maillard-type reactions from carbohydrates and amino acids, remain mainly unknown. One reason for that lack of information is that most studies focused primarily on the quantitatively predominating compounds, rather than selecting the target compounds to be identified with regard to taste-activity. As the counterpart to the aroma extract dilution analysis in aroma research, no comparative efficient screening procedures are available, which will enable a straightforward identification of taste compounds in complex food fractions.

In order to bridge the gap between pure structural chemistry and human taste perception, the aim of the present investigation was, therefore, (i) to rank the taste compounds formed during thermal treatment of carbohydrates and primary amino acids in their taste impacts by application of an activity-guided screening procedure, and then (ii) to estimate their contribution to the overall taste of the Maillard mixture on the basis of the taste activity concept.

Experimental

Thermally treated xylose/alanine mixture

A solution of D-xylose (75 mmol) and L-alanine (19 mmol) in phosphate buffer (50 mL; 0.1 mol/L, pH 5.0) was heated under reflux for 3h, then extracted with ethyl acetate (8 × 20 mL). After removing the volatiles in vacuo, this extract was used for performing the taste dilution analysis (TDA).

Taste dilution analysis (TDA)

The extract was separated by HPLC to give 21 fractions which were collected in glass vials. After freeze-drying, serial 1+1 dilutions of each fraction in water were presented in order of increasing concentrations to trained panellists, while wearing a nose clamp, and each dilution was sensorially judged in a triangle test. The dilution at which a taste difference between the diluted fraction and two blanks could just be detected, was defined as the Taste Dilution (TD) factor (*10*).

^{13}C NMR detected [^{13}C] labeling experiment

A solution of D-ribose (63.3 mmol), [$^{13}C_5$]-D-ribose (3.3 mmol) and L-alanine (33.3 mmol) in phosphate buffer (45 mL; 1 mmol/L, pH 3.0) was refluxed for 20 min, then, furan-2-aldehyde (25.0 mmol) was added, and heating was continued for another 3h. After cooling, the bitter compounds **I** (1.9 mg) and **II** (1.0 mg) were isolated by HPLC with a purity of > 99 %, and were then taken up in DMSO-d_6 for a ^{13}C NMR experiment (*11*).

Synthetic preparation of 1-oxo-2,3-dihydro-1*H*-indolizinium-6-olates I-VII

A solution of xylose (0.8 mol), L-alanine (0.2 mol) in phosphate buffer (500 mL; 0.1 mol/L, pH 5.0) was heated under reflux for 20 min, then either a binary mixture of furan-2-aldehyde (0.06 mol) and 5-methyl-furan-2-aldehyde (0.1 mmol), 5-hydroxymethyl-furan-2-aldehyde (0.05 mol) or thiophen-2-aldehyde (0.05 mol), respectively, was added and heating was continued for 150 min. After cooling, the target compounds **I-VII** were isolated by means of high-speed countercurrent chromatography and column chromatography (*11, 12*).

Results and Discussion

Sensory analysis of an aqueous solution of the solvent-extractables isolated from a heated xylose/alanine mixture revealed an intense bitter taste of the reaction products formed. To characterize the tastants, mainly evoking the bitter taste of that processed carbohydrate/amino acid solution, we introduced the so called Taste Activity Concept, consisting of the following four analytical steps:

- Localisation of the most intense taste compounds by application of an activity-guided screening procedure (Taste Dilution Analysis).
- Identification of the compounds evaluated with the highest taste impacts.
- Quantification of the most taste-active compounds identified.
- Calculation of taste activity values on the basis of a dose/activity relationship and estimation of the taste contribution.

Screening for the key players in bitter taste

Aimed at identifying the key tastants, first, the odor-active volatiles were removed by high-vacuum distillation, and the non-volatile reaction products were then screened for taste activity using an activity-guided fractionation procedure combining chemical/instrumental techniques and human sensory analysis as displayed in Figure 1.

To rank the non-volatile reaction products in their relative taste impact, we developed an HPLC-assisted bioassay, which we call Taste Dilution Analysis (TDA). To achieve this, an aliquot of the non-volatile fraction was separated by HPLC (Figure 2, left side), the effluent was separated into 21 fractions, which were collected separately and freeze-dried (*10*).

The 21 fractions collected were then stepwise 1:1 diluted with water and then presented in order of increasing concentrations to trained sensory panellists, who were asked to evaluate the taste quality and to determine the detection threshold in a triangle test. As the dilution at which a taste difference between the diluted fraction and two blanks could just be detected is proportional to the taste activity of the individual fractions in water, this so-called Taste Dilution (TD)-factor ranked the 21 HPLC fractions in their relative taste intensity (Figure 2, right side). Due to the high TD factor of 512, fraction no. 19 (Figure 2, right side) was evaluated with by far the highest taste impact, therefore, mainly contributing to the bitter taste of the Maillard mixture (*10*).

Isolation and identification of bitter key compounds in fraction no. 19

To isolate sufficient amounts of the key compounds evoking the strong bitter taste of fraction no. 19, the non-volatile fraction of the Maillard mixture was separated in a preparative scale by high-speed countercurrent chromatography (HSCCC) using n-hexane/ethyl acetate/water as a ternary

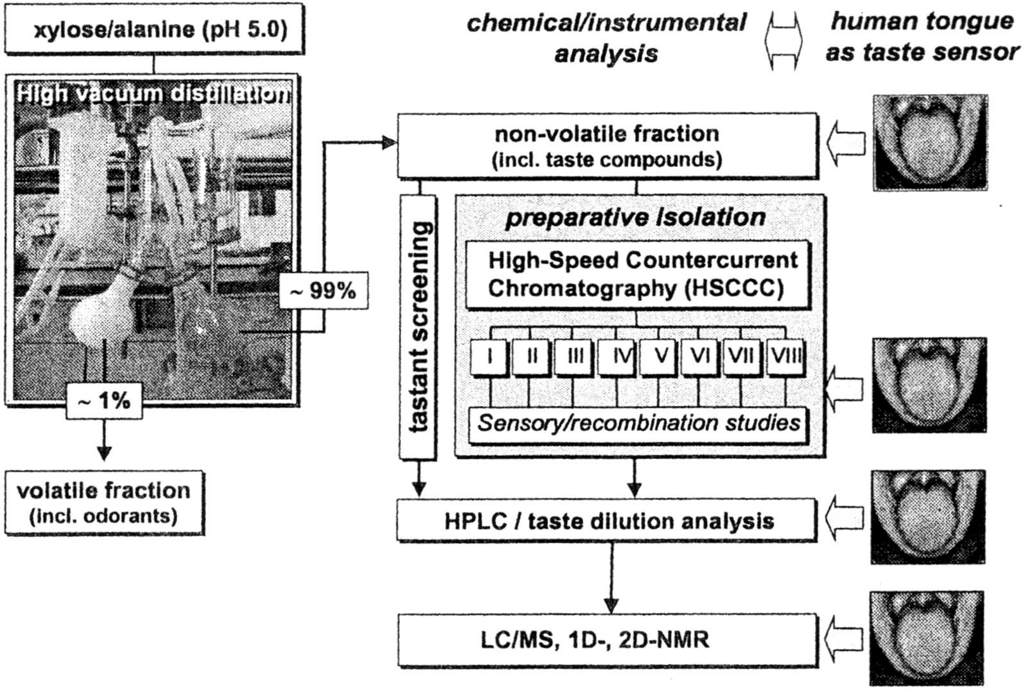

Figure 1. Screening for key taste compounds in the xylose/L-alanine solution by means of an activity-guided fractionation procedure

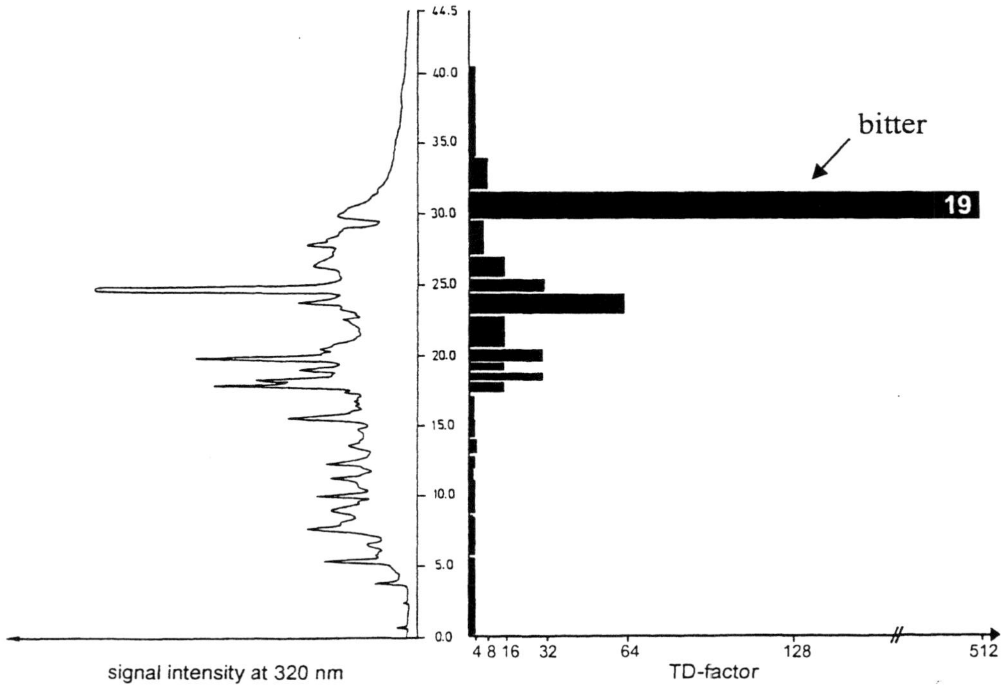

Figure 2. RP-HPLC chromatogram (left side) and Taste Dilution (TD)-chromatogram (right side) of the non-volatile, solvent extractables of a heated xylose/L-alanine solution

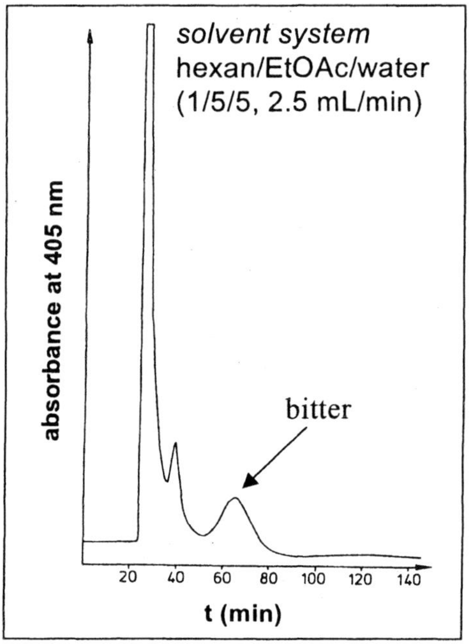

Figure 3. High-speed countercurrent chromatography (HSCCC) of the non-volatile, solvent extractables of a heated xylose/L-alanine solution

solvent system (Figure 3). Monitoring the effluent at $\lambda = 405$ nm, fractions were collected and analysed by RP-HPLC (12).

The broad peak eluting between 55 and 75 min (marked with an arrow in Figure 3), revealed two fluorescent compounds **I** and **II** showing identical UV/Vis spectra with absorption maxima at 412 and 335 nm (Figure 4). On comparison of retention times, spectroscopic as well as sensory data, **I** and **II** were confirmed to be identical with the compounds evoking the bitter taste of the HPLC fraction no. 19 in Figure 2.

After isolation by semipreparative HPLC, 1D- and 2D-NMR experiments as well as LC/MS spectroscopy were performed in order to determine the chemical structure of the bitter compounds **I** and **II**. But these identification experiments did not allow an unequivocal structure elucidation of the tastants. Neither one-dimensional, nor more sophisticated two-dimensional NMR experiments succeeded in differentiating between the two possible structures A and B (Figure 5) proposed for the major bitter compound **I** (*10, 11*).

The heteronuclear multiple bond correlation (HMBC) spectroscopy showed C,H-correlation between the hydrogen at C(4) and the carbon atom C(3) as well as connectivities between the methine proton at C(10) and the carbon atoms C(2), C(3), C(4), and C(11), respectively, fitting well with both proposed

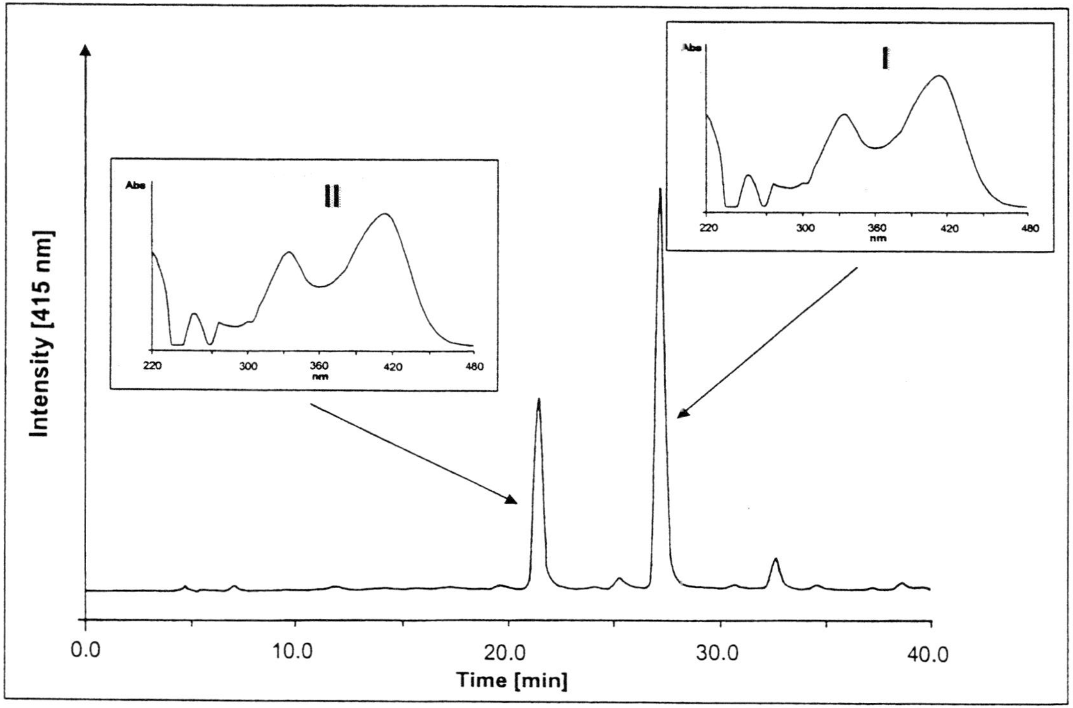

Figure 4. RP-HPLC of bitter tasting HSCCC fraction collected between 55 and 75 min (marked with an arrow in Figure 3)

structures, A and B in Figure 5. It was, therefore, not possible to elucidate whether the methine carbon C(10) in **I** is part of a six-membered ring as proposed in structure A, or represents the exocyclic methylidene bridge between the furan ring C(11)-C(14) and the heterocyclic, five-membered ring as proposed in structure B (Figure 5).

Because LC/MS and NMR spectroscopic measurements alone did not enable the unequivocal structure determination of **I**, the following considerations aimed at receiving additional structural informations by understanding via which carbohydrate fragments the tastant **I** is formed from pentoses. It was very likely from the proposed structures A and B (Figure 5) that both the furan rings correspond to two molecules of furan-2-aldehyde, a well-known major degradation product formed from pentoses. Assuming that the intact C_5-carbon skeleton of furan-2-aldehyde is incorporated into the taste compounds, it has to be accepted that, besides both the furan rings, also the neighbouring carbon atoms C(3) and C(15) in structures A and B (Figure 5) originate from furan-2-aldehyde. This indicates that the furan-2-aldehyde derived carbon atom C(3) would interrupt the contiguity of the carbon skeleton in the 1-oxo-1*H*,4*H*-quinolizinium-7-olate ring system in structure A (Figure 5), whereas in structure B (Figure 5) the contiguity of the carbon skeleton in the five-membered ring would not be touched. The visualization of the furan-2-aldehyde incorporation into **I** would, therefore, open the possibility to differentiate between structure A and B and, in consequence, to elucidate the correct structure of the tastant.

To achieve this, a first experiment focused on proving the key role of furan-2-aldehyde as one of the precursors involved in the formation of **I**. Recent quantitative model studies on aqueous solutions of xylose and L-alanine, which were thermally treated in the absence or presence of furan-2-aldehyde, revealed that the yield of **I** inceased by a factor of six when the Maillard reaction was performed in the presence of furan-2-aldehyde (*11*), clearly demonstrating that furan-2-aldehyde acts as an effective precursor for compound **I** (*11*).

In order to visualize the incorporation of the precursors into the structure of **I**, a second experiment aimed at monitoring the flux of intact carbon skeletons from the pentose into compound **I** by means of a carefully planned [^{13}C] labeling strategy. As the joint transfer of several ^{13}C atoms *en bloc* from a multiply ^{13}C labeled precursor documents that the bonds between the respective atoms have remained intact during its transformation, the site-specific detection of intact carbon bonds by measuring the ^{13}C/^{13}C coupling constants using ^{13}C NMR is a powerful tool to visualize the flux of reaction intermediates into the target compound **I**. Because compound **I** is ultimately derived from the complex pool of transient reaction intermediates such as, e.g. furan-2-aldehyde and deoxyosones etc., its labeling pattern must reflect the carbon skeletons of intermediates involved in their formation. To visualize the mosaics assembled from [^{13}C] and [^{12}C] labeled carbon modules, the [$^{13}C_5$] carbohydrate was, therefore, diluted with natural [^{13}C] abundant carbohydrate prior to the Maillard reaction (*11*).

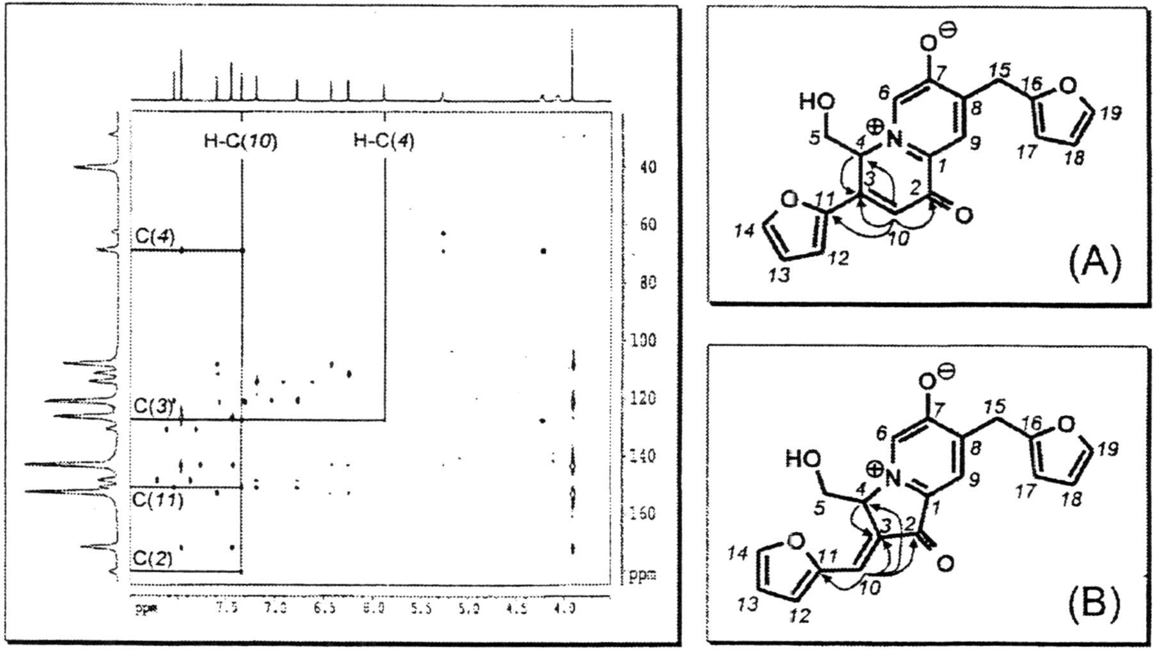

Figure 5. HMBC experiment (500 MHz, DMSO-d_6) of the purified bitter compound I and proposed chemical structures A and B

To unequivocally confirm the structure of tastant **I** and to elucidate its formation pathway, this so-called "bond labeling" technique was performed by heating natural [^{13}C] abundant ribose, which was diluted with 5.0% [$^{13}C_5$] ribose, in the presence of L-alanine in aqueous solution. In order to black-out the $^{13}C/^{13}C$ couplings in the furan-2-aldehyd-derived fragments of the compound **I**, the mixture was spiked with natural [^{13}C] abundant furan-2-aldehyde prior to thermal treatment. After solvent extraction, the isotopomeric mixture of **I** was isolated by RP-HPLC, and then analyzed by ^{13}C NMR spectroscopy. The chemical shifts and the $^{13}C/^{13}C$ coupling patterns measured for each individual carbon atom of the heterocyclic core of **I** are displayed in Figure 6 (*11*).

Due to their natural [^{13}C] abundance, the ten carbon atoms C(15)-C(19) and C(10)-C(14) were detected as singlets, thus confirming that the intact C_5 skeleton of the furan-2-aldehyd, which was added as the natural [^{13}C] abundant compound, is incorporated into **I** (data not shown). In contrast, the carbon atom C(6) resonating at 129.9 ppm showed intense ^{13}C satellites as a double doublets with coupling constants of 62.4 and 15.5 Hz corresponding to the $^1J_{C,C}$ coupling with C(7) and the $^2J_{C,C}$ coupling with C(8). For carbon C(7) a strongly coupled signal pattern was observed at 171.0 showing coupling with the carbon atoms C(6), C(8) and C(9), respectively (Figure 6). Also for carbon atom C(8) a complex coupling pattern was observed. This quaternary carbon atom shows coupling with C(9), C(7), and C(6), but did not show any connectivity with C(15), thus confirming that C(15) is not ^{13}C enriched. The atom C(9) showed coupling with C(7) and C(8), but did not with C(1). Taking all these data into account, it can be concluded that the carbon module C(6)-C(9) is shifted into the pyridinium ring of **I** as an intact C_4 fragment derived from the pentose.

To clarify the situation in the lower ring of **I**, the coupling patterns of C(1)-C(5) were investigated (Figure 6). The hydroxymethyl group C(5) resonating at 62.0 ppm showed only one homonuclear coupling to C(4), thus indicating that C(4) is indeed the only carbon atom directly connected with C(5) and confirms the finding that the C(5) is a terminal carbon atom in the second carbon module. In addition, for C(1) the coupling constants were found to be $^1J_{C(1)/C(2)}$ = 67.4 and $^2J_{C(1)/C(3)}$ = 17.2 Hz, but no further coupling was observable to C(9) and C(10). These data unequivocally confirmed the presence of the exocyclic (2-furyl) methylidene group as proposed for structure B of compound **I** in Figure 5. Being well in line with structure B, the complex signal patterns of the carbon atoms C(2) or C(3), respectively, showed coupling with C(1), C(3) and C(4), or C(1), C(2) and C(4) as illustrated in Figure 6. On the other hand, the lack of coupling between C(2) and C(10) or C(3) and C(10), respectively, collaborated with the singlet detected for C(10) at 117.8 ppm (Figure 6).

Considering all the ^{13}C-NMR data obtained for the ^{13}C enriched tastant **I**, it can be concluded that the quaternary carbon atom C(3) in structure B (Figure 5) is part of the intact C_5 carbon module showing the original contiguity of the carbon skeleton of the pentose, and does not originate from the carbonyl carbon of the precursor furan-2-aldehyde. On the basis of these data, the correct structure of compound **I** could be unequivocally elucidated to be the previously not reported (2*E*)-7-(2-furylmethyl)-2-(2-furylmethylidene)-3-(hydroxymethyl)-

1-oxo-2,3-dihydro-1*H*-indolizinium-6-olate (**I** in Figure 7), called quinizolate (*11*).

Analogous ^{13}C labeling experiments were performed to elucidate the chemical structure of the bitter compound **II** (Figure 4) detected in HPLC fraction no. 19 (Figure 2). Assignment of the $^{13}C/^{13}C$ correlations in the ^{13}C-enriched compound **II** showed major similarities in chemical shifts and $J_{C/C}$ coupling constants for the carbon atoms C(1)-C(5) and C(6)-C(8), but differed strongly in the multiplicity of C(9). The carbon C(9) showed connectivities to C(7), C(8), and, in addition, to an additional hydroxymethyl group. All spectroscopic data obtained for tastant **II** unequivocally revealed its chemical structure as (2*E*)-7-(2-furylmethyl)-2-(2-furylmethylidene)-3,8-bis(hydroxymethyl)-1-oxo-2,3-dihydro-1*H*-indolizinium-6-olate (**II** in Figure 7), called homoquinizolate, which to the best of our knowledge was not reported earlier in the literature (*11*).

Quantification and calculation of taste activity values

To determine the taste contribution of these compounds more accuratly, **I** and **II** were quantified in the Maillard reaction mixture by HPLC/DAD using the pure reference compounds as external standards. As given in Table I, the quinizolate is formed in more than two-fold higher concentrations as homoquinizolate, e.g. 0.167 and 0.077 mmol/kg of **I** and **II**, respectively, have been determined.

Table I. Concentration, detection thresholds and taste activity values (TAV) of quinizolate (I) and homoquinizolate (II), and their contribution to the overall bitter taste of the heated xylose/alanine mixture

Tastant[a]	*Concentration [mmol/kg]*	*Detection threshold*[b] *[mmol/kg water]*	*TAV*[c]	*Contribution to total bitterness [%]*[e]
I	0.167	0.00025	668	65
II	0.077	0.001	77	8
	Σ (TAV_I+TAV_{II}):		745	73.0
	TD_{total}-factor[d]:		1024	100.0

[a] The structure of the tastants is displayed in Figure 7.

[b] The detection threshold was determined in water using a triangle test.

[c] The taste activity value (TAV) was calculated from the ratio of the concentration and the detection threshold of a tastant.

[d] The taste activity of the overall Maillard mixture was determined by measuring the taste dilution factor.

[e] The taste contribution was calculated as follows:
contribution (%) = (TAV / TD_{total} factor) $\times$ 100

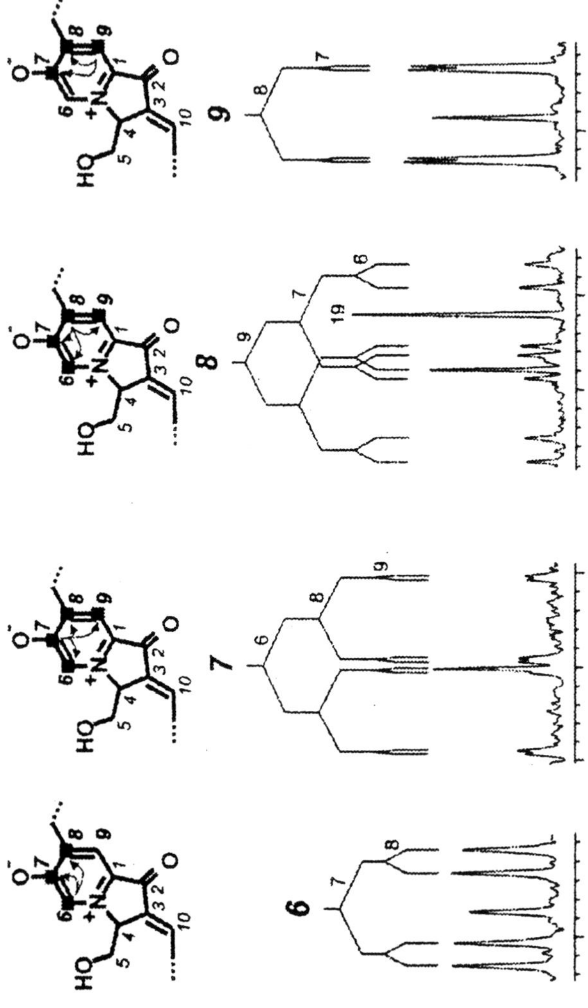
6
7
8
9
130.0
171.0
142.5
142.0
120.5
ppm

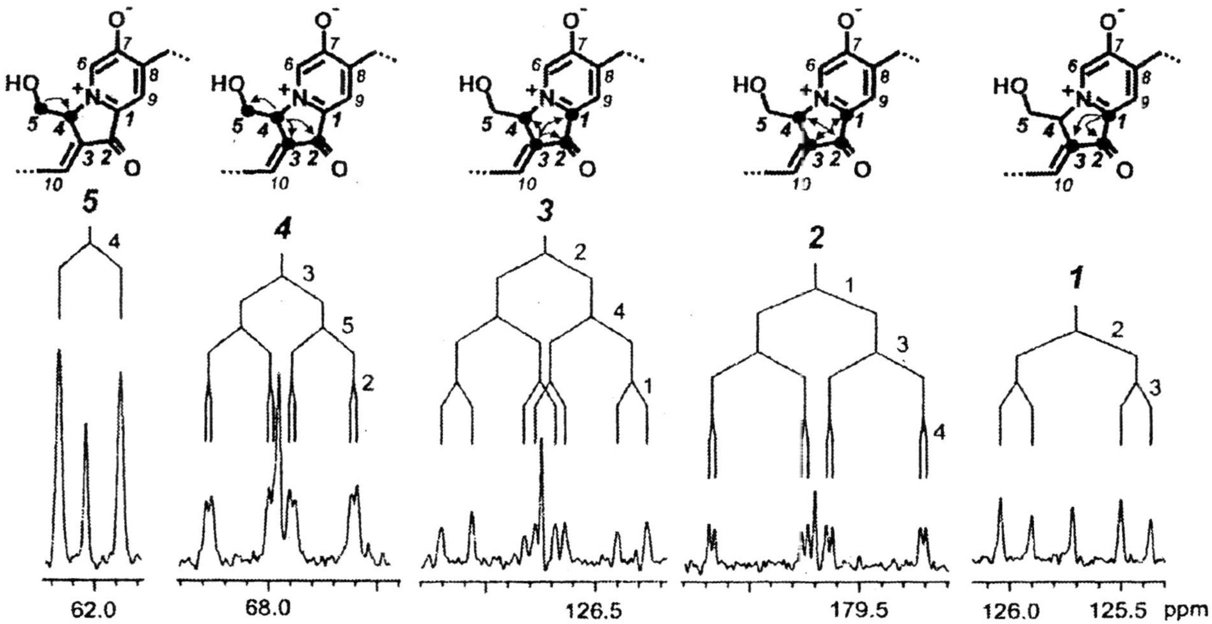

Figure 6. Excerpts of ^{13}C NMR spectrum of compound I isolated from a heated ribose/alanine/furan-2-aldehyde mixture containing 5% [$^{13}C_5$] ribose. Arbitrary numbering of carbon atoms refer to structure B in Figure 5.

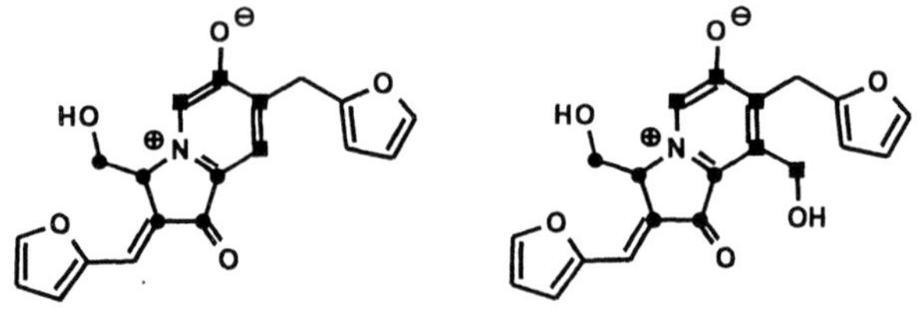

I II

Figure 7. Structures and [^{13}C] labelling patterns of quinizolate (I) and homoquinizolate (II). "■" and "●" should differentiate ^{13}C atoms originating from different pentose molecules.

Since the quantitative data alone do not allow an estimation of the importance of the tastants **I** and **II** in evoking the overall bitter taste of the xylose/alanine solution, their taste activity values (TAVs) were calculated from the quotient of the actual concentration and the human taste detection thresholds. To achieve this, first, the detection thresholds of the tastants were determined using a triangle test. Quinizolate and homoquinizolate were evaluated with extraordinarily low bitter detection thresholds of 0.00025 and 0.001 mmol/kg (water), respectively (Table I). Relating the concentration with the detection threshold of the tastants then revealed a TAV of 668 for quinizolate indicating that the concentration of that tastant in the Maillard mixture is 668-fold above the threshold (Table I). In comparison, homoquinizolate was evaluated with a somewhat lower TAV of 77.

To estimate the percent taste contribution of an individual compound, the measurement of the overall bitterness of the Maillard solution is a necessary prerequisite. To achieve this, the taste dilution factor of the complete reaction mixture, the TD_{total}-factor, was determined. The complete Maillard solution was, therefore, diluted with water until no bitter taste difference between a diluted aliquot and two blanks containing tap water could be detected using a triangle test. A TD_{total}-factor of 1024 was found for the total mixture indicating that the taste of the undiluted, original reaction mixture was 1024-fold above its detection threshold. Also, the TAV of a tastant corresponds, by definition, to the factor to which the actual concentration is above the detection threshold. The percentual taste contribution of a single tastant could, therefore, be estimated from the TDA of the compound and the TD_{total} factor of the complete Maillard mixture, which was defined as possessing a taste activity of 100 % (Table I). The TAV of 668 for **I** means that the actual concentration of **I** is 668-fold above its detection threshold (Table I). Since the complete Maillard mixture, accounting for 100 % of the bitterness, is 1024-fold above its detection threshold, it can be estimated that about 65 % of the overall bitterness is caused by **I**. In summary, **I** and **II**, contributed to 73 % to the total bitterness of the reaction mixture (Table 1).

Formation pathways leading to 1-oxo-2,3-dihydro-1*H*-indolizinium-6-olates

Besides the verification of the chemical structures, the [^{13}C] labeling experiments also allowed insights into the reaction pathways how tastants **I** and **II** are formed from pentoses (Figure 8). Amine-assisted water elimination from the pentose (1) liberates the 3,4-dideoxypentosulose (2), a major dehydration product formed from pentoses (*13*), which upon Strecker-type reactions with primary amino acids leads to the amino ketone 3 by reductive amination. Subsequent cyclization by an intramolecular Michael-type reaction yields the 5-hydroxymethyl-3-oxo-pyrrolidine (4), which undergo further condensation with a molecule of 3,4-dideoxypentosulose (2) giving rise to the transient iminium ion 5. Reaction between the two CH-acidic methylene groups in 5 and furan-2-aldehyde results in the formation of the bis-condensation product 6. This intermediate acts as a chemical switch, which has the possibility to undergo either a retro-Aldol cleavage of one molecule of formaldehyde, or a simple imine/enamine tautomerism. Retro-Aldol cleavage of formaldehyde generates the intermediate 7, which upon aromatization give rise to quinizolate (**I**). As an alternative, formation of the transient enamine 8, followed by aromatization results in homoquinizolate (**II**).

Structure and sensory activity of 1-oxo-2,3-dihydro-1*H*-indolizinium-6-olates

To gain first insights into the relationship between the chemical structure and the human psychobiological activity of quinizolate (**I**) and homoquinizolate (**II**), five additional 1-oxo-2,3-dihydro-1*H*-indolizinium-6-olates (**III-VII** in Figure 9) varying in the substitution of the furan moiety were synthesized (*13*).

Utilizing the finding that heating pentoses and primary amino acids in the presence of furan-2-aldehyde led to the incorporation of the aldehyde into **I**, xylose and alanine were reacted in the presence of either 5-methylfuran-2-aldehyde, 5-hydroxymethyl-furan-2-aldehyde, or thiophen-2-aldehyde, and the target compounds **III-VII** (Figure 9), were isolated by means of high-speed countercurrent chromatography as well as column chromatography. After final HPLC purification, the chemical structures of compounds **III-VII** were confirmed by LC/MS, NMR and UV/Vis spectroscopy, and the human bitter taste thresholds were determined by means of a triangle test with a trained sensory panel.

The highest taste activity was evaluated for the thiophen derivative **VII**, which showed the extraordinarily low bitter detection threshold of 6.3×10^{-5} mmol/kg (water), followed by quinizolate (**I**) showing a four-fold higher detection threshold concentration (*13*). The incorporation of a methyl group into the (2-furan)methylidene system in compound **III** did not infuence the bitter threshold of the 1-oxo-2,3-dihydro-1*H*-indolizinium-6-olate (Figure 9), whereas

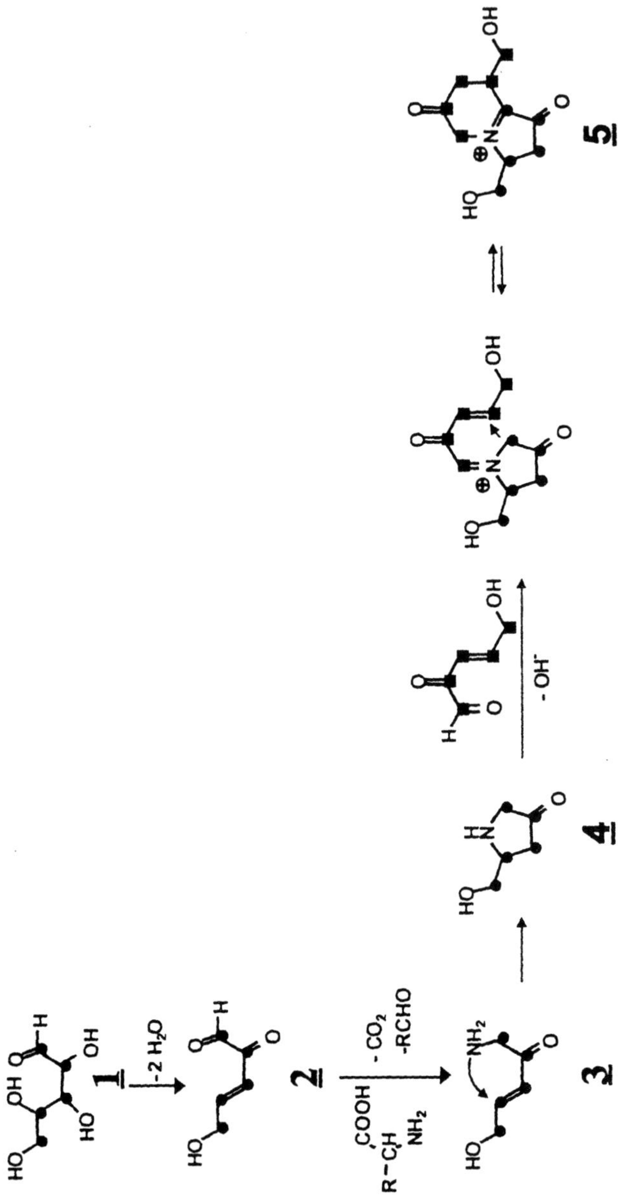
-2 H2O
-CO2
-RCHO
-OH-
1
2
3
4
5

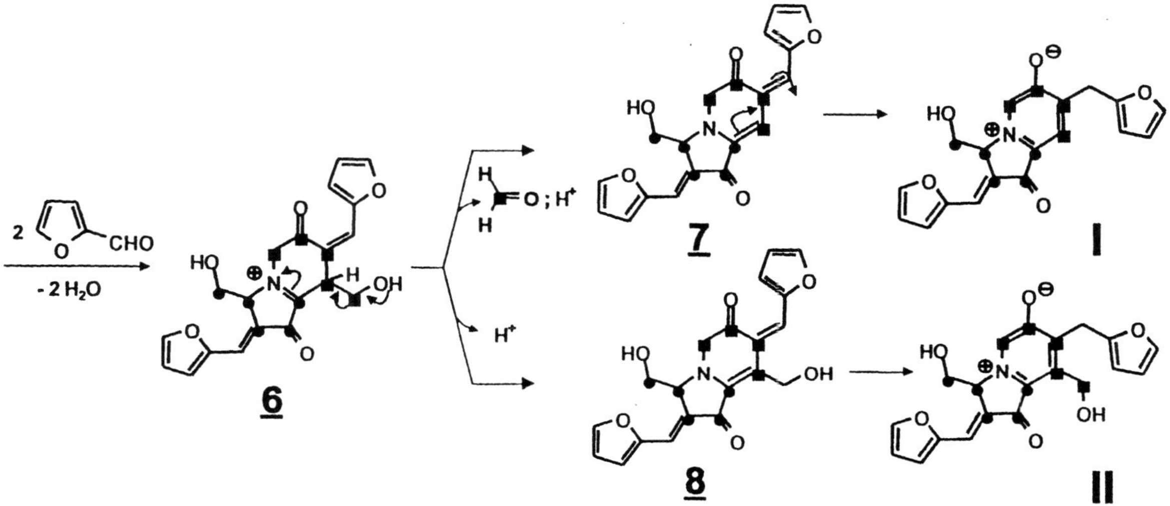

Figure 8. Reaction pathway leading to the formation of quinizolate (I) and homoquinizolate (II) from pentoses ("■" and "●" should differentiate ^{13}C atoms originating from different pentose molecules in the labelling experiments).

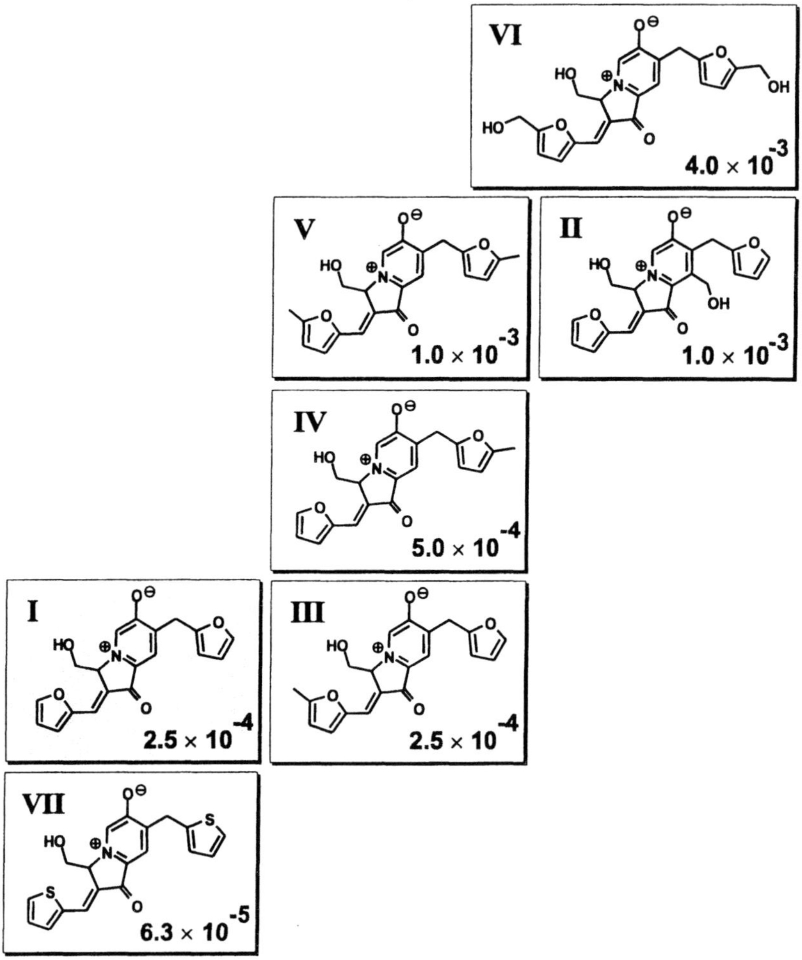

Figure 9. Chemical structures and human bitter thresholds (mmol/kg water) of 1-oxo-2,3-dihydro-1H-indolizinium-6-olates I-VII.

the presence of a methyl group in the upper furan ring of compound **IV** led to an increase of the detection threshold by a factor of two. The substitution of both furan rings in **I** by a 5-methyl-2-furyl moiety increased the threshold four-fold to 1.0 × 10^{-3} mmol/kg (water). The same threshold was determined for homoquinizolate (**II**) bearing an additional hydroxymethl group at the pyridinium ring. The attachment of two hydroxymethyl groups to the furan rings of **I** induced a 6-fold increase in the taste threshold, e.g. a threshold concentration of 4.0 × 10^{-3} mmol/kg (water) was determined for **VI** (*13*). Taking all these data into consideration, it might be concluded that the incorporation of polar groups such as, e.g. the hydroxymethyl groups, led to a pronounced increase in detection threshold. Additional methyl groups at the furan rings of quinizolate (**I**) led to a somewhat lower effect on threshold increase. In contrast, the substitution of the oxygen atoms in the furan rings of **I** by sulfur atoms induced a significant decrease of the detection threshold of the 1-oxo-2,3-dihydro-1*H*-indolizinium-6-olate. Most likely, the more "voluminous" sulfur atoms in **VII** are favored by the corresponding taste receptor in comparison to the oxygen analogon **I** (*13*).

Conclusions

The taste dilution analysis (TDA) has been demonstrated as a activity-guided screening technique enabling a straightforward localization of tongue-affecting compounds in complex mixtures of compounds present in Maillard reaction mixtures or foods. The taste activity concept comprising the identification of the key tastants, followed by quantification and estimation of their taste contribution on the basis of taste activity values, reveals useful informations on the key players evoking the typical taste of foods.

References

(1) Day, E.A.; Forss, D.A.; Patton, S. *J. Dairy Sci.* **1957**, *40*, 932-941.
(2) Acree, T.E.; Bernard, J.; Cunningham, D.G. *Food Chem.* **1984**, *14*, 273-286.
(3) Schieberle, P.; Grosch, W. *Z. Lebensm. Unters. Forsch.* **1987**, *185*, 111-113.
(4) Schieberle, P.; Grosch, W. *Z. Lebensm. Unters. Forsch.* **1994**, *198*, 292-296.
(5) Schieberle, P. *Food Chem.* **1996**, *55(2)*, 145-152.
(6) Schieberle, P.; Hofmann, T. *J. Agric. Food Chem.* **1997**, *45*, 227-232.
(7) Buettner, A.; Schieberle, P. *J. Agric. Food Chem.* **2001**, *49*, 2387-2394.
(8) Warmke, R.; Belitz, H.D.; Grosch, W. *Z. Lebensm. Unters. Forsch.* **1996**, *203*, 230-235.

(9) Schlichterle-Cerny, H.; Grosch, W. *Z. Lebensm. Unters. Forsch.* **1998**, *207*, 369-376.
(10) Frank, O.; Ottinger, H.; Hofmann, T. *J. Agric. Food Chem.* **2001**, *49*, 231-238.
(11) Frank, O.; Hofmann, T. *J. Agric. Food Chem.* **2002**, *50*, 6027-6036.
(12) Frank, O.; Jezussek, M.; Hofmann, T. *Eur. Food Res. Technol.* **2001**, *213*, 1-7.
(13) Frank, O.; Hofmann, T. Structure and sensory activity of bitter 1-oxo-2,3-dihydro-1*H*-indolizinium-6-olates. *J. Agric. Food Chem.*, in prep.

Chapter 8

Chemistry of Theaflavins: The Astringent Taste Compounds of Black Tea

Chi-Tang Ho, Shengmin Sang, and Jin Woo Jhoo

Department of Food Science, Rutgers, The State University of New Jersey, 65 Dudley Road, New Brunswick, NJ 08901–8520

Tea is one of the most popular beverages consumed worldwide. Freshly harvested tea leaves require processing to convert them into green, oolong and black teas. An estimated 2.5 million metric tons of dried tea are manufactured annually. Here, we review the chemistry of theaflavins, the astringent taste compounds in black tea. In the course of studies on the oxidation mechanism of tea polyphenols, a new type tea pigment, theadibenzotropolone A, together with theaflavin 3-gallate were formed by the reaction of (-)-epicatechin (EC) and (-)-epigallocatechin gallate (EGCG) with horseradish peroxidase in the presence of H_2O_2. The structure of theadibenzotropolone A was elucidated on the basis of MS and 2D NMR spectroscopic analyses. A compound containing two types of benzotropolone links is of great interest, since it shows that the galloyl ester group of theaflavin monogallate participates in further oxidative condensation reactions. This may provide an additional reaction pathway for the formation of thearubigins. The existence of this compound in black tea was characterized by LC/ESI-MS/MS.

Tea, a beverage produced by brewing the dried leaves and buds of the plant *Camellia sinensis*, has been consumed by humans for thousands of years. *Camellia sinesis* was first cultivated in China and then in Japan. With the opening of ocean routes to the East by European traders during the fifteenth to seventeenth centuries, commercial cultivation gradually expanded to Indonesia and then to the Indian subcontinent, including what is now Sri Lanka (*1*). Tea is now second only to water in worldwide consumption. Annual production of about 1.8 million tons of dried leaf provides world per capita consumption of 40 liters of beverages (*2*).

Based on different processing methods, green tea, black tea, and oolong tea are the three major types of tea products. Tea polyphenols, also knows as catechins, account for 30% to 42% of water-soluble solids in brewed green tea. There are four major tea catechins: (−)-epigallocatechin-3-gallate (EGCG), (−)-epigallocatechin (EGC), (−)-epicatechin (EC) and (−)-epicatechin gallate (ECG) (*2*). The structures of these catechins are shown in Figure 1. These catechins, especially EGCG, have been reported to have various biological activities (*3,4*). Tea consumption has been suggested to protect against cancer, and cardiovascular and other diseases, but the results from epidemiological studies are not conclusive (*3-6*).

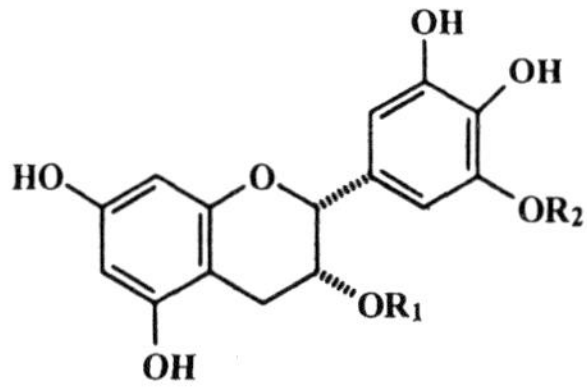

		R_1	R_2
(-)-Epicatechin	EC	H	H
(-)-Epicatechin gallate	ECG	Gallate	H
(-)-Epigallocatechin	EGC	H	Gallate
(-)-Epigallocatechin gallate	EGCG	Gallate	Gallate

Figure 1. Structures of catechins in teas.

Commercial Tea Processing

Tea leaf, like all other plant leaf matter, contains carbohydrate, protein, lipids, a full complement of genetic material, enzymes and secondary metabolites. In addition, tea leaf is distinguished by its high content of methylxanthins and polyphenols, in particular catechins. Table 1 shows a representative analysis of fresh tea leaf. 25% of its dry weight are catechins, 3.0% are flavonols and flavonol glycosides, 3.0% are caffeine and 0.2% are theobromine (*2*).

In the processing of green tea, the tea flesh is first steamed in the case of Japanese green tea "sen-cha" or pan-fired to produce Chinese green tea (*6*). These heat treatments inactivate enzymes in the tea leaves. The temperature of pan-firing can reach as high as 230°C which is much higher than the steaming temperature of 100°C. Steaming, therefore, results in fewer chemical changes than pan-firing. Following heat treatment, tea leaves are subjected to subsequent rolling and drying processes to achieve a dry product exhibiting the desired twisted leaf appearance.

Table I. Composition of Fresh Tea Leaf

Components	*% of dry weight*
Polyphenols	39
Methyl xanthines	3.5
Amino acids	4.0
Organic acids	1.5
Carbohydrates	25
Protein	15
Lignin	6.5
Lipids	2
Chlorophyll and other pigments	0.5
Ash	5

SOURCE: Reproduced from Reference 2. Copyright 1992, American Chemical Society.

More than 75% of world tea production is black tea. The steps involved in the processing of black tea include withering, leaf disruption, fermentation, drying and grading. All steps are designed to achieve optimal oxidation of tea catechins and produce tea products with good flavor and color.

Oolong tea is prepared by firing the leaves after rolling to terminate the oxidation process. It is only partially oxidized and retains a considerable amount of the original polyphenols (*7*).

Chemical Changes During Black Tea Processing

During black tea manufacturing, tea leaves are first subjected to a withering process until moisture contents reach to ~75-80%. Withered leaves are then ruptured through a rolling process. The polyphenols in tea leaf are located in the vacuole, however, polyphenol oxidase, which is responsible for fermentation, is localized in the chloroplasts. Shearing stresses of the rolling process enable polyphenols to diffuse freely into the cytoplasm and initiate enzymatic oxidation. The rolling process is not, however, long enough for fermentation to take place, so the leaves are heaped either on sheet metal racks or on the floor of the fermenting room for about 40 min to 3 hours for further fermentation. When fermentation is completed, the leaves are fired. The firing process is responsible, not only for inactivating the enzymes, but also for reducing the moisture content (~2.5-3.5%) of the tea (*8*).

Most of the important chemical changes take place during the fermentation process. Polyphenol oxidase, which is responsible for oxidation of flavonols, oxidizes pyrogallol and catechol to their *o*-quinones, and further chemical reactions make various oxidation products. During fermentation, characteristic black tea pigments are generated and those pigments are divided into two groups, named theaflavin and thearubigin. Four major theaflavins have been identified from black tea, including theaflavin, theaflavin-3-gallate and theaflavin-3′-gallate, theaflavin-3,3′-digallate. The structures of these major theaflavins are shown in Figure 2. Thearubigin is known as a heterogeneous mixture of pigments.

Roberts et al. (*9*) reported fermented black tea contains polyphenolic pigments, which are not found in the unprocessed tea leaves. They designated the brown acidic pigments and the yellow neutral pigments as thearubigins and theaflavins, respectively. They suggested that theaflavin was coupling oxidation products of EGC and EGCG having benzotropolone structure. Later, Takino et al. (*10*) corrected the structure of theaflavin, and revealed that theaflavin was coupling oxidation products of EC and EGC. Theaflavin formation was confirmed by enzymatic oxidation with crude tea polyphenol oxidase and chemical oxidation with potassium ferricyanide including its structure determination using NMR technique. It was clear that theaflavins are produced by enzymatic oxidation of di-hydroxylated (catechol) and tri-hydroxylated B ring (pyrogallol) to their quinones, and follow by condensation.

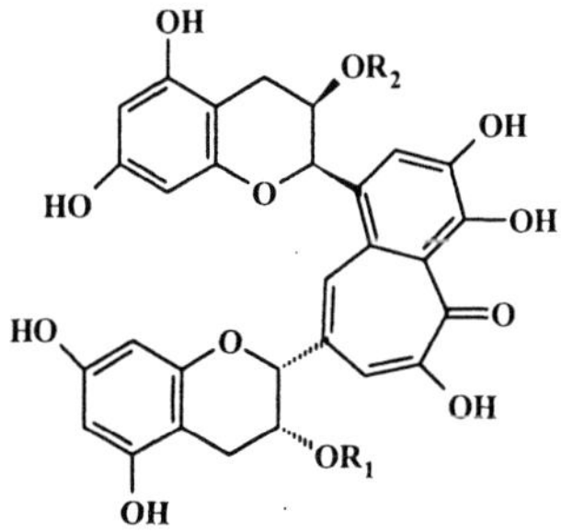

		R_1	R_2
Theaflavin	TF	H	H
Theaflavin 3-gallate	TF3G	Gallate	H
Theaflavin 3'-gallate	TF3'G	H	Gallate
Theaflavin 3,3'-gallate	TFDG	Gallate	Gallate

Figure 2. Structures of theaflavins.

Takino et al. (*10*) proposed a theaflavin formation mechanism based on formation of purpurogallin from pyrogallol (Figure 3). The results suggested the possibility of the existence of similar pigments from the other pairs of flavonols (Table II). Bryce et al. (*11*) successfully isolated three other pigments from black tea, and confirmed their parent flavonols using chemical oxidation methods. One of them, named TF1, was identical to the theaflavin reported by Roberts (*9*). TF2 was identified as a mixture of TF2A and TF2B, which were produced by chemical oxidation from EGCG and EC (TF2A) and EGC and ECG (TF2B), respectively. TF3 was an identical compound, which was isolated from ferricyanide oxidation of EGCG and ECG. They identified TF4 as an oxidation product from gallic acid and EC. Later, Collier et al. (*12*) confirmed the presence of four theaflavins in black tea, and also confirmed the presence of epitheaflavic acid and epitheaflavic acid gallate, which are identical oxidative products from the coupling of EC and gallic acid, and ECG and gallic acid, respectively. Isotheaflavin, which is reported by Coxon et al. (*13,14*) was also identified as present in black tea, and was confirmed as an oxidative coupling product of EGC and catechin. Compared to theaflavins, epitheaflavic acid, theaflavic acid and epitheaflavic acid gallate occur at a much lower level. The relative proportions of the theaflavins in black tea were theaflavin (18%), theaflavin-3-gallate (18%), theaflavin-3'-gallate (20%), theaflavin-3,3'-digallate (40%), and the proportions of theaflavic acids, along with isotheaflavin, were approximately 4% (*8*). Several other minor pigments have also been reported from black tea, theaflavate A, theaflavate B, isotheaflavin-3'-O-gallate and neotheaflavin-3-*O*-gallate (*15,16*).

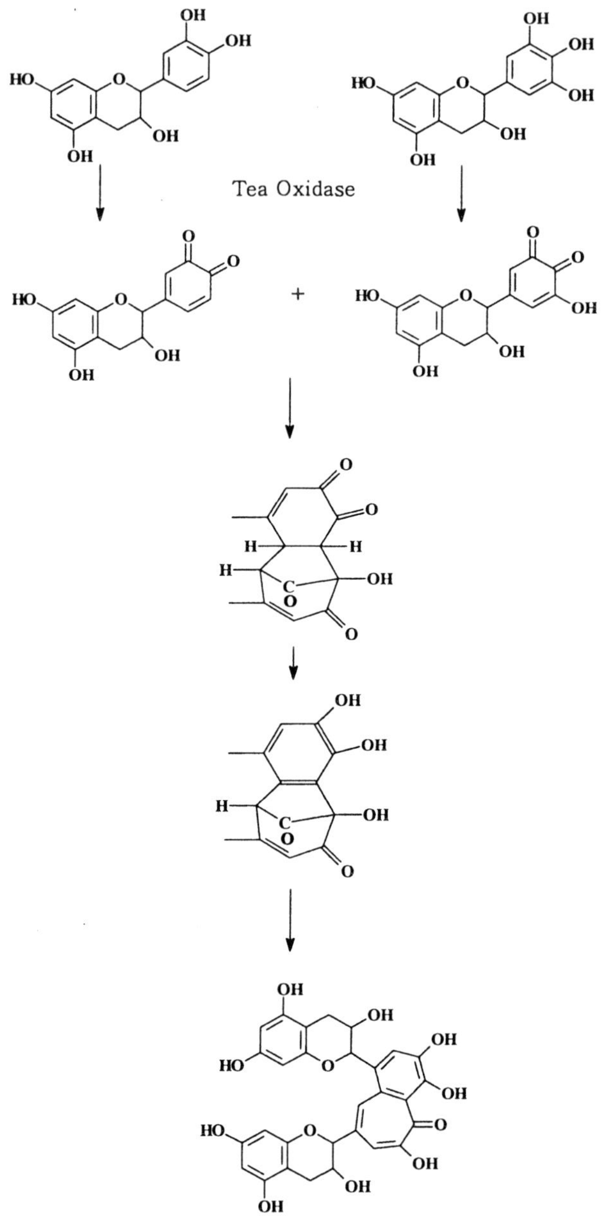

Figure 3. The proposed mechanism of theaflavin formation (Modified from Reference 10)

Thearubigins consist of ~10-20% of dry weight of black tea. However, because of its hot water solubility they account for ~30-60% of the solids in black tea infusion (*9*). Even though Roberts suggested the thearubigins are further oxidation products of theaflavins and catechins, their formation mechanism and chemical structures are not clear. Thearubigins have a wide range of molecular weights, ~700-40,000, and are regarded as polymer compounds. Brown et al. (*17*) reported that thearubigins are a polymeric mixture of proanthocyanidin containing flavonoids residue, and they proposed its formation mechanism through creation of a C-C bond. Later, Berkowitz et al. (*18*) suggested that formation of epitheaflavic acid from EC and gallic acid plays an important role for formation of thearubigins. They found that when EC was added into a tea fermentation system with epitheaflavic acid, both compounds disappeared rapidly, and thearubigins content increased. Interestingly, when it was reacted with only epitheaflavic acid in a tea fermentation system no reaction occurred. However, the exact chemistry of thearubigins remains unclear.

Table II. Parent Flavonols of Theaflavins and Epitheaflavic Acids

Theaflavins	*Parent flavonols*
Theaflavin	(-)-epicatechin and (-)-epigallocatechin
Theaflavin-3-gallate	(-)-epicatechin and (-)-epigallocatechin gallate
Theaflavin-3′-gallate	(-)-epicatechin gallate and (-)-epigallocatechin
Theaflavin-3,3′-digallate	(-)-epicatechin gallate and (-)-epigallocatechin gallate
Epitheaflavic acid	(-)-epicatechin and gallic acid
Epitheaflavic acid gallate	(-)-epicatechin gallate and gallic acid

Theaflavins and the Astringent Taste of Black Tea

Flavor is the most important factor in determining the quality of tea. Flavor involves both taste and aroma. The balance of astringency, bitterness and brothy taste is important to the characteristic taste of tea (*19*). The major contributors to astringency and bitterness of tea are catechins and caffeine.

Several studies revealed a relationship between chemical composition of tea leaves and final quality of black tea products (*20*). During fermentation process, catechin contents of tea leaves are decreased, whereas, new oxidation products, such as theaflavins and thearubigin, are increased. However, due to differences in redox potential of catechins, tea catechins are depleted at different rates. For example, pyrogallol (EGC, EGCG) may deplete faster than catechol (EC, ECG). Interestingly, catechins have a different degree of astringency. Table III lists the threshold levels for astringency and bitterness of catechins, theaflavins and caffeine.

In general, gallated catechins have more astringency. The theaflavins have astringent tastes and contribute to the briskness of black tea. Theaflavin-3,3′-digallate is the most astringent (6.4 times higher than theaflavin), and theaflavin monogallates rank next to digallate for astringency. Thus, theaflavins composition is regarded as an important factor affecting the color and briskness of black tea. Generally, extended fermentation time and increased fermentation temperature is not suitable to obtain high quality of final black tea products. Robertson et al. (*21*) reported that, according to *in vitro* oxidation experiments, high oxygen tension, low pH (pH 5.0) and low temperature (30°C) factors gave high yield of theaflavins formation. Black tea fermentation is a complicated mechanism that can influence the final quality of black tea, not only by catechins composition of tea leaves, but also by polyphenol oxidase activity.

However, from Table III, it is clear that the concentration of catechins and four major theaflavins can not account for the perceived astringent taste of black tea beverages. It is, therefore, our effort to search for other, higher molecular weight theaflavins presented in black tea.

Model Studied on the Reaction of EC and EGCG and the Identification of Theadibenzotropolone A in Black Tea

In order to better understand the enzymatic formation of theaflavins, model studies were performed on the reaction of EC and EGCG. EC (**1**) and EGCG (**2**) were treated with horseradish peroxidase in the presence of H_2O_2. In addition to theaflavin 3-gallate (**3**), a new type tea pigment, theadibenzotropolone **A** (**4**), was isolated.

The molecular formula of **4** was determined to be $C_{50}H_{38}O_{21}$ by negative-ion ESI-MS ($[M-H]^-$ at *m/z* 973) as well as from its ^{13}C NMR data, which indicates

Table III. Threshold Levels for Astringency and Bitterness of Catechins and Theaflavins in Black Tea

Polyphenol	*Threshold Level (mg/100 mL)*		*Approximate Level in a Cup of Black Tea* (mg/100 mL) of Beverage*
	Astringency	*Bitterness*	
(-)-Epicatechin	Not astringent	60	Trace
(-)-Epicatechin gallate	50	20	Trace
(-)-Epigallocatechin	Not astringent	35	Trace
(-)-Epigallocatechin gallate	60	30	16-18
(+)-Catechin	Not astringent	60	Trace
Theaflavin	80	75-100	0.6-1.2
Natural mixture of theaflavin monogallate	36	30-50	1.8-3.7
Theaflavin 3,3'-digallate	12.5	unknown	2.4-4.8
Gallic acid	Not astringent	Not bitter	3-5

- Based on amount of solids extracted from a standard American tea bag (2.27 g tea) brewed with 6 oz boiling tap water in a cup for 3 min.

that this compound consists of three flavan-3-ol units. This was confirmed by ^{1}H- and ^{13}C- NMR data. The ^{13}C-NMR spectrum of **4** displayed 50 carbon signals, 27 of which were assigned to the A and C rings of flavan-3-ols. In addition, the ^{1}H-NMR spectrum exhibited three sets of signals due to protons at the 2-, 3-, 4-, 6-, and 8- positions of the flavan-3-ol nucleus. These observations also indicate that the A and C rings of **4** did not undergo any change during oxidation. In comparison with the ^{1}H-NMR spectrum of **3**, compound **4** is distinguished by the absence of galloyl ester signals, a large downfield shift of H-3′, an additional set of A and C ring signals from flavan-3-ol, and three more olefinic proton signals (δ 7.54 brs H-c; 7.79 s H-g; 8.30 brs H-e). In the ^{13}C-NMR spectrum of **4**, besides the A and C ring signals, there were observed 23 carbon signals including two carbonyls (δ 186.5 s C-a; 185.5 s C-a′), one ester carbonyl (δ 167.7 s C-l) and 20 olefinic carbons. All of these spectral features support the presence of two benzotropolone groups in compound **4**. Thus, the galloyl ester group on **3** can react with the B-ring of EC to form another benzotropolone. This assertion is supported by the 2D HMBC NMR spectrum. Thus, the structure of **4** was deduced as shown (Figure 4) and named theadibenzotropolone **A**.

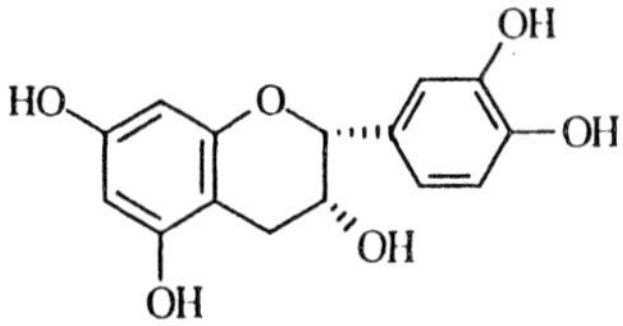

1

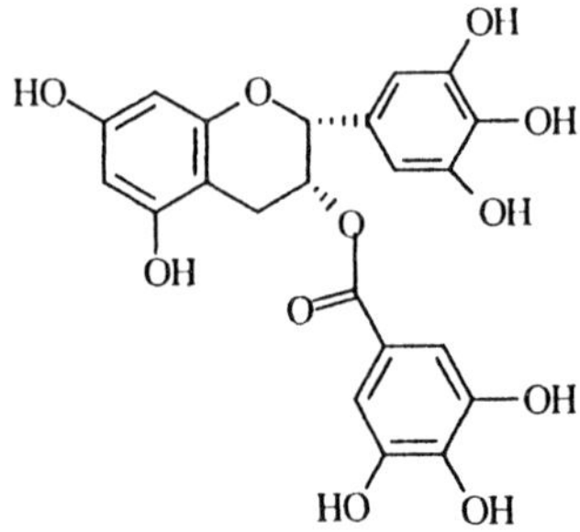

2

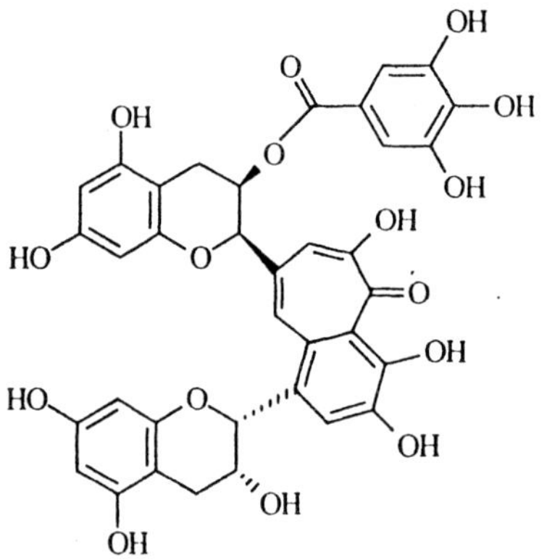

3

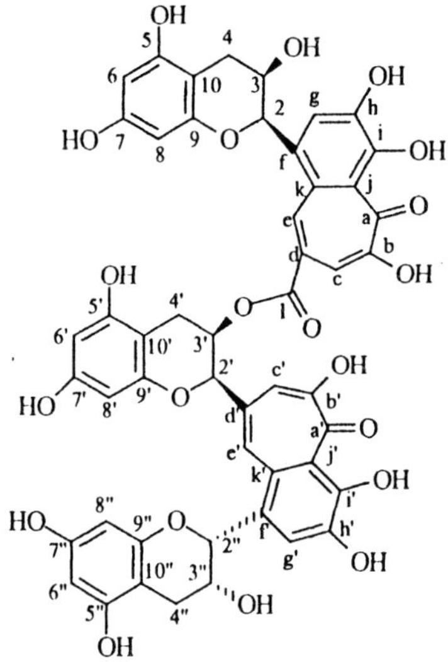

4

Figure 4. Structures of compounds 1-4.

Efforts were made to establish the presence of theadibenzotropolone **A** in black tea using selected-ion monitoring (SIM) chromatograms of the standard (compound **4**) and the theaflavin fraction of black tea isolated from the 80% acetone aqueous extraction of black tea by Sephadex LH-20 column chromatography, respectively. Figure 5 shows the respective LC/MS/MS chromatograms and fragment ion mass spectra of compound **4** and the theaflavin fraction of black tea. As shown in this figure, compound **4** and the peak in the theaflavin fraction of black tea exhibited not only the same chromatographic retention time (tR) and molecular masses, but also had the same fragment ion mass spectra. This proves the presence of theadibenzotropolone **A** in black tea.

It is well known that the major oxidizing enzyme in tea is polyphenol oxidase. It is the action of this enzyme that mainly produces the theaflavins. So the formation, in black tea, of theadibenzotropolone **A** may come from the reaction between theaflavin 3-gallate and the free catechin EC by the action of peroxidase in the presence of H_2O_2. This was confirmed by the reaction of theaflavin 3-gallate and EC in the same conditions as that of EC and EGCG. All of the theaflavin 3-gallate can change to theadibenzotropolone **A** if there is enough free EC. Thus, the synthesis of theadibenzotropolone **A** via enzymatic oxidation of EC (**1**) and EGCG (**2**) and EC and theaflavin 3-gallate (**3**) is highly significant. Theadibenzotropolone **A** provides the first demonstrable proof that the galloyl ester group of theaflavins is as reactive as the B-ring (*vic*-trihydroxy) of EGCG or EGC (epigallocatechin) and the galloyl ester group of ECG (epicatechin gallate). The observation that galloyl ester groups of theaflavins can be oxidized to form dibenzotropolone skeletons strongly implicates that this type of oxidation as an important pathway to extend the molecular size of thearubigins.

In the LC/MS/MS study on black tea, we found that there are at least three peaks corresponding to the molecular weight of m/z 974. We are currently synthesizing the possible isomers of theadibenzotropolone **A** and will try to identify them in black tea. The contribution of these higher molecular weight theaflavin-like compounds to the astringent and bitter taste of black tea will be evaluated.

References

1. Forrest, D. *The World Trade: A Survey of the Production, Distribution and Consumption of Tea*; Woodhead-faulkner, Ltd.: Cambridge, UK, 1985.Yang, C.S.; Chung, J.Y.; Yang, G.; Chhabra, S.K.; Lee, M.J. *J. Nutr.* **2000**, *130 (2S Suppl)*, 472S-478S.
2. Balentine, D. A. In *Phenolic Compounds in Food and Their Effects on Health I. Analysis, Occurrence, & Chemistry*; Ho, C.-T.; Lee, C.Y.; Huang, M.T., Eds.; American Chemical Society: Washington. D.CRC Press: Boca Raton, FL, 1998; pp. 35-72.
3. Yang, C.S.; Maliakal, P.; Meng, X. *Annu. Rev. Pharmacol. Toxicol.* **2002**, *42*, 25-54.
4. Blot, W.J.; Chow, W.H.; McLaughlin, J.K. *Eur. J. Cancer Prev.* **1996**, *5*, 425-438.
5. Yang, C.S.; Yang, G.Y.; Chung, J.Y.; Lee, M.J.; Li, C. *Adv. Exp. Med. Biol.* **2001**, *492*, 39-53.
6. Yamanishi, T.; Kawakami, M.; Kobayashi, A.; Hamada, T.; Musalam, Y. In *Thermal Generation of Aromas*; Parliment, T. H.; McGorrin, R. J.; Ho, C.-

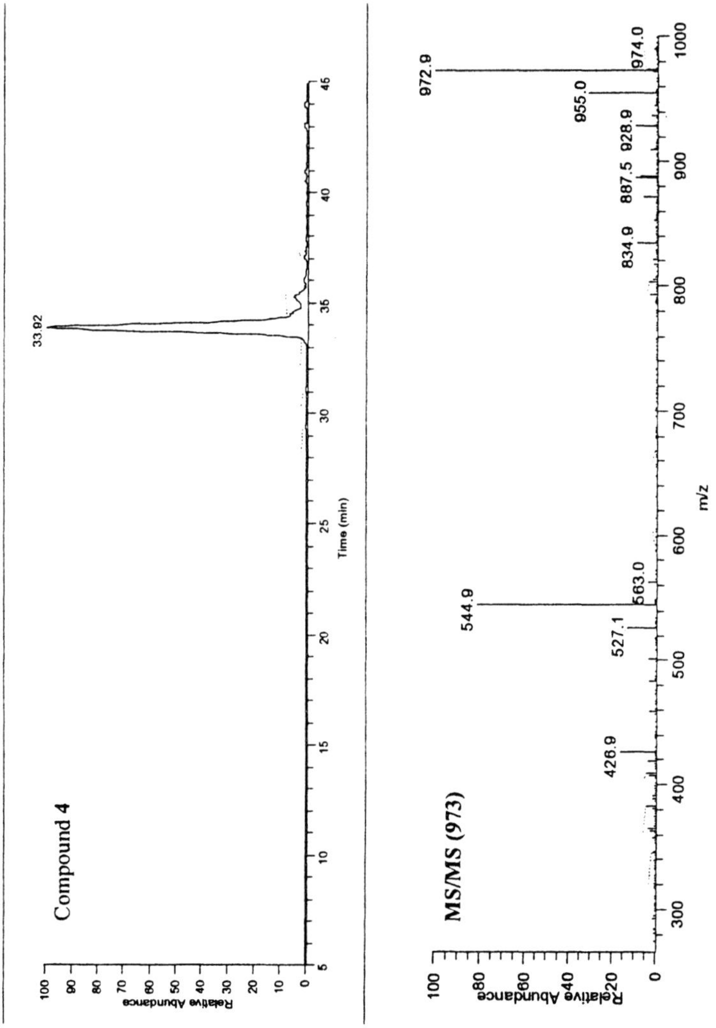
Compound 4
33.92
Time (min)
Relative Abundance
MS/MS (973)
972.9
974.0
955.0
928.9
887.5
834.9
563.0
544.9
527.1
426.9
m/z
Relative Abundance

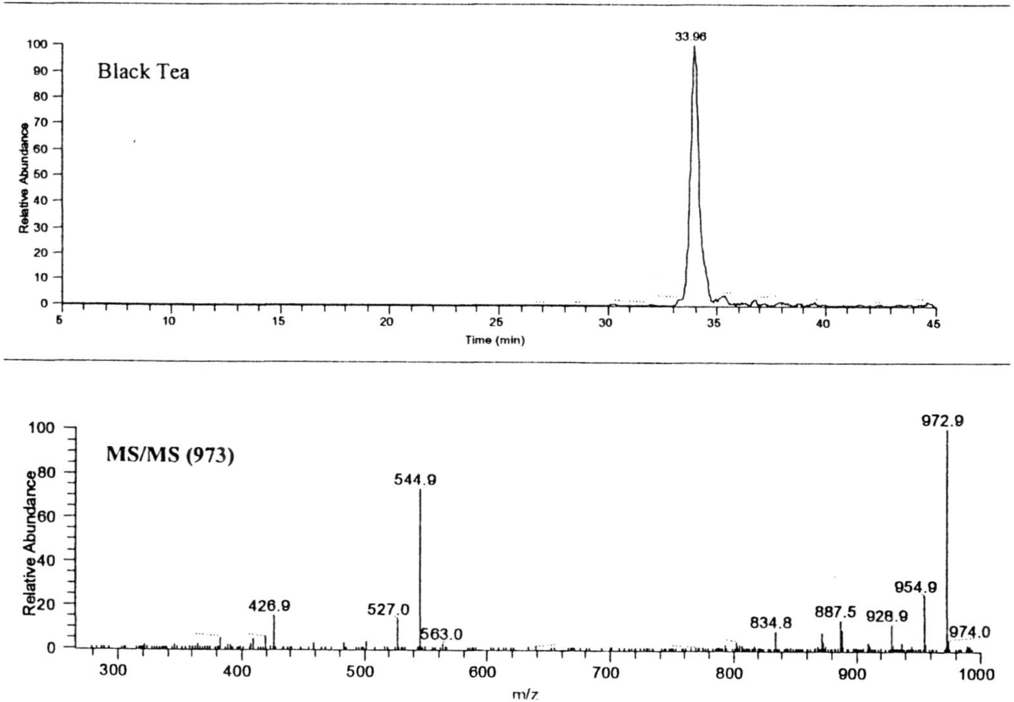

*Figure 5. Negative ion LC/ESI-MS/MS chromatograms and fragment ion mass spectra of compound **4** (Top) and the peak in the theaflavin fraction of black tea (Bottom).*

T.; ACS Symposium Series 409; American Chemical Society: Washington, DC, 1992; pp. 102-117.

7. Graham, H. N. In *The Methylxanthine Beverages and Foods: Chemistry, Consumption and Health Effects*; Spiller, G. A., Ed.; Alan R. Liss: New York, 1984; pp. 29-74.
8. Robertson A. In *Tea, Cultivation to Consumption*, Chapman & Hall: London, UK. 1992; 555-601.
9. Roberts, E.A.H.; Cartwright, R.A.; Oldschool, M. *J. Sci. Food Agric.* **1957**, *8*, 72-80.
10. Takino, Y.; Imagawa, H.; Horikawa, H.; Tanaka, A. *Agric. Biol. Chem.* **1964**, *28*, 64-71.
11. Bryce, T.; Collier, P.D.; Fowlis, I.; Thomas, P.E.; Frost, D.; Wilkins, C.K. *Tetrahedron letter* **1970**, *32*, 2789-2792.
12. Collier, P.D.; Mallows, B.R.; Thomas, P.E.; Frost, D.J.; Korver, O.; Wilkins, C.K. *Tetrahedron* **1973**, *29*, 125-142.
13. Coxon, D.T.; Holmes, A.; Ollis, W.D. *Tetrahedron Letter* **1970**, *60*, 5241-5246.
14. Coxon, D.T.; Holmes, A.; Ollis, W.D.; Vora, V.C. *Tetrahedron Letter* **1970**, *60*, 5237-5240.
15. Wan, X.; Nursten, H. E.; Cai, Y.; Davis, A.L.; Wilkins, J.P.G.; Davis, A.P. *J. Sci. Food Agric.* **1997**, *74*, 401-408.
16. Lewis, J.R.; Davis, A.L.; Cai, Y.; Davies, A.P.; Wilkins, J. P. G.; Pennington M. *Phytochemistry* **1998**, *49*, 2511-2519.
17. Brown, A.G.; Eyton, W.B.; Holmes, A.; Ollis, W. D. *Phytochemistry* **1969**, *8*, 2333-2340.
18. Berkowitz, J.E.; Coggon, P.; Sanderson G.W. *Phytochemistry* **1971**, *10*, 2271-2278.
19. Yamanishi, T. *Food Rev. Internat.* **1995**, *11*, 477-525.
20. Obanda, M.; Owuro, P. O.; Mang'oka, R. *Food Chem.* **2001**, *75*, 395-404.
21. Robertson A. *Phytochemistry* **1983**, *22*, 889-896.

Chapter 9

Pungent and Tingling Compounds in Asian Cuisine

Christophe C. Galopin[1], Stefan M. Furrer[1], and Andreas Goeke[2]

[1]Ingredient Systems, Givaudan Flavors R&D, 1199 Edison Drive, Cincinnati, OH 45069
[2] Fragrance Research, Givaudan Dübendorf Ltd., Uberlandstrasse 138, 8600 Dübendorf, Switzerland

Southern Asian cuisine is well known for its use of flavorful and pungent spices. The sanshool chemicals, such as alpha-hydroxy-sanshool from the Japanese Sanchoo pepper and other Asian peppers, are particularly interesting because they not only give a hot sensation in the mouth cavity but also a tingling effect on the tongue. In order to understand the effect of the sanshool chemicals we have synthesized a variety of derivatives. Tasting of those derivatives provided information about Structure Activity Relationship (SAR) for the tingling effect exhibited by these chemicals. Based on this study we are able to propose a minimal structure required for the tingling effect. We also used this SAR knowledge to design stable compounds with potential tingling effect.

Introduction

The research on pungent compounds has been very active in the past few years thanks to commercial and scientific interests. On the commercial side, consumer trends show an increased demand for strongly pungent flavors all over the world. This demand is created by Americans and Europeans who are more and more interested in spicy Asian and Latin American foods and by the expanding Asian and Latin American markets. The classic pungent compounds, piperine and capsaicin, do not satisfy the consumers' need for "burn-to-death" pungency. On the scientific side, pungency is a very interesting area. Pungency is a mouthfeel which involves non-volatile molecules that interact with trigeminal nerves. In contrast with cooling, which involves cold thermoreceptor, pungency (or hotness) involves trigeminal pain nerves. In fact, many pungent chemicals are weak anesthetics as they saturate the pain nerve response preventing any further pain signal from being transmitted to the brain. Pungency can therefore be described as a pleasurable pain in the mouth.

Thanks to our TasteTrek™ expeditions to Asia and to our interest in Sanshoo pepper, we became interested not only in pungent chemicals but also in chemicals that create a tingling sensation on the tongue. It has been reported that long-chained amides from the sanshool and bungeanool families exhibit pungent and tingling properties and have long been used as anesthetics in folk medicine (*1*). Nevertheless, these chemicals are difficult to synthesize and are often unstable (*2*). We wondered whether we could stabilize these molecules by removing unnecessary features. Although it was noticed that one of the cis-double bond in the fatty chain was a key element of activity (*3*), little structure activity relationship (SAR) information is available. Here we wish to describe our research to identify the elements of sanshools and bungeanools that are required for pungency. We will show how that knowledge can be applied to the synthesis of new molecules with the desired properties.

Experimental

Methyl (2E,4E,8Z,11Z)-tetradeca-2,4,8,11-tetraenoate ***21***

To a solution of methyl 4-(diethoxyphosphoryl)crotonate (4.66g (19.7 mmol) in THF (50 ml) was added ^{t}BuOK (2.40g, 19.7 mmol) at 0°C and the mixture was stirred for 30 minutes. The red-brown solution was cooled to –30°C and (4Z,7Z)-decadienal **20**[1] (1.90g, 13.2 mmol) was added dropwise. The mixture was warmed to room temperature during 30 minutes and poured into a saturated solution of NH_4Cl. The mixture was extracted with MTBE, the organic phase was washed with water and brine, dried ($MgSO_4$) and concentrated *in vacuo*. The yellow residue was purified by chromatography to yield 1.31g (45%) of a yellowish oil (containing about 10 % of the 4Z-isomer). ^{1}H-NMR (400MHz, $CDCl_3$) (δ in ppm): 7.26 (dd, J = 15.6, 10.0 Hz, 1H), 6.21-6.08 (m, 2H), 5.80 (d, J = 15.6 Hz, 1H), 5.44-5.25 (m, 4H), 3.74 (s, 3H), 2.77 (dd, J = 7, 7 Hz, 2H), 2.25-2.15 (m, 4H), 2.07 (dq, J = 7, 7 Hz, 2H), 0.97 (t, J = 7 Hz, 3H); ^{13}C-NMR

(100MHz, $CDCl_3$): 167.6 (s), 145.1 (d), 143.8 (d), 132.0 (d), 129.1 (d), 128.7 (d), 128.4 (d), 126.9 (d), 119.0 (d), 51.4 (q), 32.9 (t), 26.4 (t), 25.5 (t), 20.5 (t), 14.2 (q); MS (EI): 234 (M^+, 1), 129 (2), 203 (3), 175 (14), 152 (17), 133 (10), 119 (12), 105 (14), 93 (15), 67 (100), 59 (9), 55 (26), 41 (15); IR (neat): 3011w, 2962m, 1719s, 1644m, 1435m, 1251s, 1135s, 999s, 723m cm^{-1}.

Bungeanool 7

Ester **21** (2.00g, 8.5 mmol) was dissolved in water/MeOH (30 ml, 1:3) and saponified with KOH (1.67g, 30 mmol) overnight. The mixture was brought to acidic pH with ice-cold HCl and extracted with MTBE. The organic phase was washed with water and brine, dried ($MgSO_4$) and concentrated *in vacuo* to yield 1.52g of the (2*E*,4*E*,8*Z*,11*Z*)-tetradeca-2,4,8,11-tetraenoic acid. This was dissolved in CH_2Cl_2 (30 ml) containing a drop of DMF. Oxalyl chloride (5.08g, 40 mmol) was added dropwise and the mixture was stirred for 8h at room temperature. The solvent and the excess of oxalylchloride were evaporated and the residue was dried *in vacuo*. The oily material was again taken up in CH_2Cl_2 and 1-amino-2-methyl-2-propanol (1.25g, 14 mmol) was added. The mixture was stirred for 1h, the solvent was evaporated and the residue purified by chromatography to yield 0.98g (40%) of bungeanool **7** as a slightly yellow oil. ^{1}H-NMR (200MHz, $CDCl_3$) (δ in ppm): 7.21 (dd, J = 15 , 11 Hz, 1H), 6.33 (bt, J = 6Hz, 1H), 6.20-6.05 (m, 2H), 6.85 (d, J = 15 Hz, 1H), 5.47-5.25 (m, 4H), 3.35 (d, J = 6 Hz, 2H), 2.77 (dd, J = 7, 7 Hz, 2H), 2.25-2.16 (m, 4H), 2.07 (dq, J = 7, 7 Hz, 2H), 1.23 (s, 6H), 0.97 (t, J = 7 Hz, 3H).

N-isobutyl (2E,4E,8Z)-undeca-2,4,8-trienamide ***28***

At 0°C, in a round-bottom flask under an inert atmosphere of nitrogen, a solution of 2.77g (11mmol) of diethyl diethyl N-isopropyl phosphonoacetamide in 20mL of dry tetrahydrofuran is added to 15.5mL of a 1.5M solution of butyl lithium (23mmol) in hexane. The mixture is stirred at 0°C for thirty minutes. A solution of 1.4g of E2,Z6-nonadienal in 5mL of dry tetrahydrofuran is then added dropwise to the stirred reaction mixture. The mixture is stirred at 0°C for two hours. The reaction mixture is then diluted in 100mL of hexane and washed with a saturated aqueous solution of ammonium chloride. The organic phase is collected and dried over anhydrous magnesium sulfate, filtered and concentrated. The residue is purified by chromatography on silica gel with EtOAc/Hexane:2/8 as the eluent to give 0.5g of product as a white fluffy powder. ^{1}H NMR (300MHz in $CDCl_3$) (δ in ppm): 7.2 (dd, J=15, 10.2 Hz, 1H), 6.1 (m, 2H), 5.75 (d, J=15 Hz, 1H), 5.34 (m, 3H), 3.2 (t, J=6.5Hz ,2H), 2.2 (m,

2H), 2.0 (quintuplet, J=7.5Hz, 1H), 1.8 (septuplet, J=6.6Hz, 1H), 0.96(t, J=7.5Hz, 3H), 0.93 (d, J=6.9Hz, 3H).

Methyl (2E,4E,8Z)-deca-2,4,8-trienoate ***30***

To a solution of methyl 4-(diethoxyphosphoryl)crotonate (11.6 g, 49.2 mmol) in THF (70 ml) was added ^{t}BuOK (5.98g, 49.2 mmol) at 0°C. The mixture was cooled to –78°C and a solution of (*Z*)-hex-4-enal (4.00g, 40.8 mmol) in THF (10 ml) was added dropwise. After the cooling bath was removed and the mixture had warmed up to room temperature, sat. NH_4Cl was added and the mixture was extracted with pentane. The organic phase was washed with water and brine, dried ($MgSO_4$) and concentrated *in vacuo*. The residue was distilled bulb to bulb (90°C/0.01 Torr) to yield **30** (2.4g, 33%) as a 7/3 mixture of the (4*E*/*Z*)-isomers. ^{1}H-NMR (200MHz, $CDCl_3$) (δ in ppm): 7.69-7.20 (m, 1H), 6.29-5.31 (m, 5H), 3.76, 3.74 (2s, 3H), 2.45-2.13 (m, 4H), 1.61 (d, J = 6.5 Hz, 3H); MS (EI): 180 (M^+, 2), 149 (6), 121 (10), 111 (27), 93 (28), 67 (51), 59 (32), 55 (100), 39 (35), 29 (28); IR (neat): 3015m, 2949m, 1720s, 1644m, 1435m, 1264s, 1137m cm^{-1}.

N-(2-hydroxy-2-methyl-propyl) (2E,4E,8Z)-deca-2,4,8-trienamide ***29***

Ester **30** (1.90g, 10.56 mmol) was saponified with NaOH (2.11g, 52.8 mmol) in H_2O/MeOH (5/1, 60 ml) for 2 days. The crude reaction mixture was brought to pH = 1 with HCl (2N) and extracted 5 times with MTBE. The organic phase was washed with brine, dried ($MgSO_4$) and concentrated *in vacuo*. The residue was dissolved in CH_2Cl_2 (30 ml) containing a drop of DMF and treated overnight with oxalyl chloride (2.0g, 15.7 mmol). The solvent was removed in vacuo (while keeping the temperature around 20°C), the residue was redissolved in CH_2Cl_2 (10 ml) and added to a solution of 1-amino-2-methyl-2-propanol (1.1g, 12.4 mmol) and triethyl amine (1.5g, 15 mmol) in CH_2Cl_2 (20 ml). The mixture was stirred for 5h and was then quenched with water and extracted with CH_2Cl_2. The organic phase was washed with 1N HCl, water and brine, dried ($MgSO_4$) and concentrated *in vacuo*. *N*-(2-hydroxy-2-methyl-propyl) (2*E*,8*Z*)-deca-2,4,8-trienamide **29** (2.0g, 80%) crystallized from ethyl acetate/hexane as a 7/3 mixture of the (4*E*/*Z*)-isomers in form of slightly yellow crystals. (4*E*)-Isomer: ^{1}H-NMR (400MHz, $CDCl_3$) (δ in ppm): 7.19 (dd, J = 15.0 Hz, 10.3 Hz, 1H), 6.56 (broad t, J = 6.0 Hz, 1H), 6.19-6.03 (m, 2H), 5.86 (d, J = 15.0 Hz, 1H), 5.52-5.44 (m, 1H), 5.39-5.32 (m, 1H), 3.49 (s, 1H), 3.33 (d, J = 6.0 Hz, 2H), 2.25-2.13 (m, 4H), 1.60 (d, J = 6.5 Hz, 3H); MS (EI): 237 (M^+, 4), 179 (62), 164 (13), 149 (23), 124 (48), 110 (100), 94 (22), 84 (25), 66 (30), 59 (47), 55

(88), 41 (24), 30 (34); IR (neat): 3287br(OH), 2974m, 2931m, 1658s, 1627s, 1609s, 1537s, 1179m, 1161m, 996m, 913m cm^{-1}.

Example of the synthesis of a cinnamamide.

In a round-bottom flask under an inert atmosphere of nitrogen, cinnamic acid (37.04g, 0.25 mol) and thionyl chloride (44.6g, 0.375mol) were dissolved in tetrahydrofurane. Two drops of pyridine were added and the mixture was heated at reflux for 4h. The red mixture was concentrated (40°C / 125mbar) and 42.7 g of crude cinnamyl choride were recovered as a brownish oil. In a round-bottom flask under an inert atmosphere of nitrogen, cinnamyl chloride (4.1 g, 25 mmol) is dissolved in a mixture of 25ml of dry tetrahydrofuran contain and 5 ml of pyridine was added. To this solution, propylamine (1.77g, 30mmol) was added over a period of 30 minutes at room temperature. The mixture was stirred for 5h at room temperature. The reaction mixture was diluted with MTBE and extracted with water. The organic phase was washed with aqueous hydrochloric acid (1N) and brine, dried ($MgSO_4$), filtered and concentrated. The residue was crystallized from MTBE/hexane to give 4.1g of product as a yellowish fluffy powder. ^{1}H NMR (300MHz, $CDCl_3$) (δ in ppm): 7.6 (d, *J*= 15.6Hz 1H), 7.5 (m, 2H), 7.4 (m, 3H), 6.4 (d, *J*= 15.9Hz, 1H), 5.8 (broad s, 1H), 3.4 (q, *J*= 6.7Hz, 2H), 1.6 (sextuplet, *J*= 7.3Hz, 2H), 1.0 (t, *J*= 7.3Hz, 3H).

Sanshool and Bungeanool Families

General description

Sanshools and bungeanools (Figure 1) are both long-chained polyenamides found in the *Achillea*, *Echinacea* and *Zanthoxylum* plant species. It seems that α-sanshool, a.k.a. echinacein and neoherculin, was first isolated from *Echinacea angustifolia* (*4*) and *Zanthoxylum piperitum* (Sanshoo pepper) (*5*) and later from *Zanthoxylum clava-herculin* (*6*). Bungeanools were first recognized as a family when isolated from *Zanthoxylum bungeanum*[2] (*7*) although similar chemicals had been described earlier (*8*).

The main difference between sanshools (**1–6**) and bungeanools (**7–10**) is the presence of three conjugated double bonds at the end of the amide chain of sanshools whereas bungeanools only have two or less unconjugated double bonds. Moreover, sanshools are either α,β or α,β,γ,δ unsaturated amides

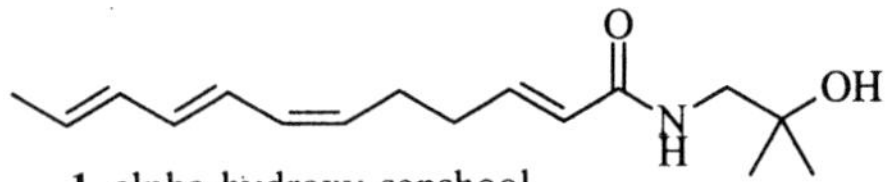

1, alpha-hydroxy-sanshool

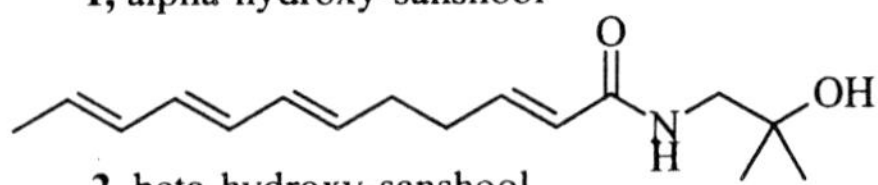

2, beta-hydroxy-sanshool

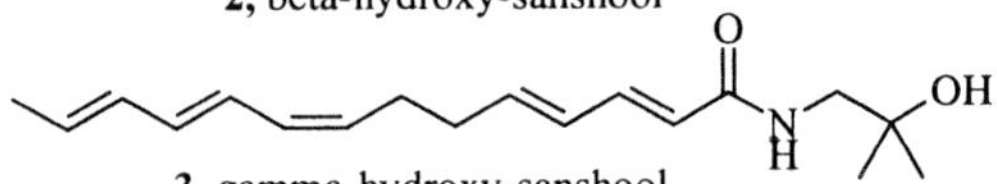

3, gamma-hydroxy-sanshool

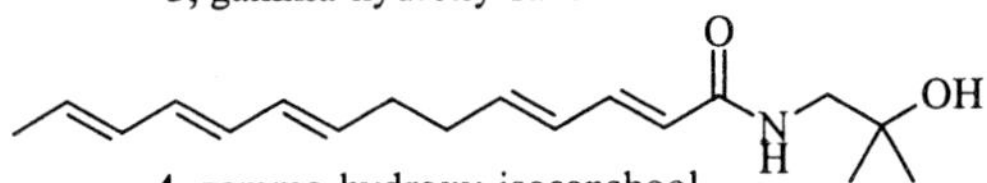

4, gamma-hydroxy-isosanshool

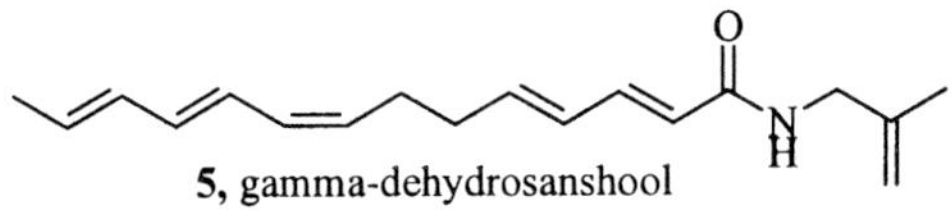

5, gamma-dehydrosanshool

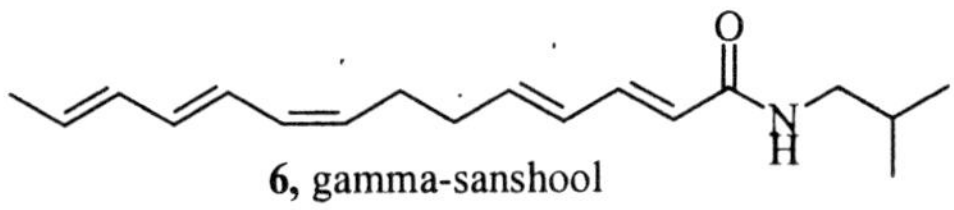

6, gamma-sanshool

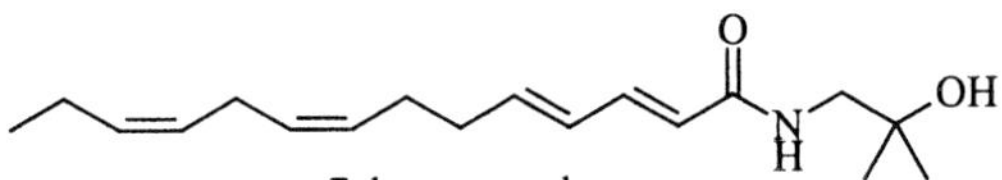

7, bungeanool

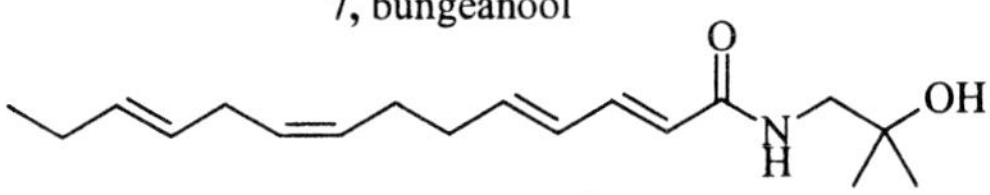

8, isobungeanool

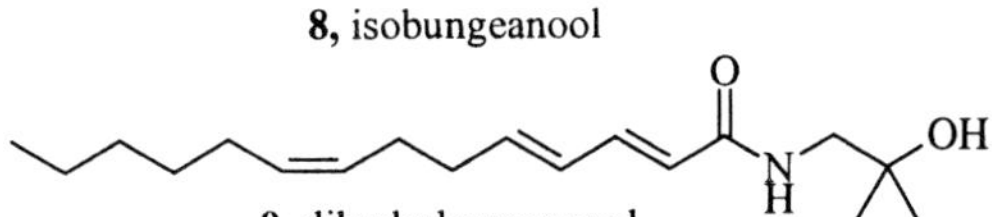

9, dihydrobungeanool

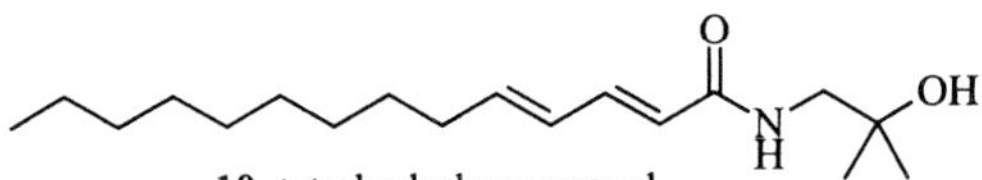

10, tetrahydrobungeanool

Figure 1 Examples of sanshools and bungeanools

whereas all bungeanools seem to be α,β,γ,δ unsaturated amides. In both families, the N-alkyl group is predominantly 2-hydroxy-2-methylpropane, but chemicals with dehydrated (**5**) or dehydroxylated (**6**) groups have been reported *7, 9*). Interestingly enough, the first sanshool isolated had no hydroxyl group on the N-alkyl chain so the word "sanshool" refers to a non-functionalized N-alkyl group whereas "bungeanool" refers to a hydroxylated N-alkyl group.

Chemical Synthesis

The difficulty of the synthesis of these chemicals depends on the ease of synthesis of the intermediate aldehydes such as **14** and **20**. These aldehydes are often expensive and complex to synthesize (*10*). Scheme 1 and 2 show examples of synthesis of α-sanshool (*2*) and bungeanool. Almost all chemicals in either family can be obtained by using similar techniques.

SAR Study of the Fatty Chain

In order to establish an SAR for these chemicals, we decided to synthesize a small series of sanshools, bungeanools and derivatives, and screen them for pungency. We decided to synthesize derivatives with a variety of functionalities representative of both families. These derivatives are α,β and α,β,γ,δ unsaturated amides with up to three unsaturations at the end of the chain. We also made some arachidonic acid derivatives (**23**–**25**) that have no conjugated double bonds.

From the pungency results shown in Figure 3, it can be concluded that having a *cis*-double in the chain is indeed a key element (*3*) but it is not the only one. The inactivity of the arachidonic derivatives show that a certain motif must be preserved. The activity of compound **22** indicates that the long chain does play a role in pungency. The activity of compound **10** indicates that the end of the chain must be unsaturated but the activity of compound **4** suggests that the unsaturation must follow a certain pattern. It becomes apparent that all active compound share a common motif (CH=Z=CH-CH2-CH2-CH=E=CH) in addition to the amide function. This motif cannot be branched (**27**) and although it seems required it is not enough to give activity (**25**). Finally, the hydroxy group on the N-alkyl group does not seem to be required since **16** and **1** are both pungent.

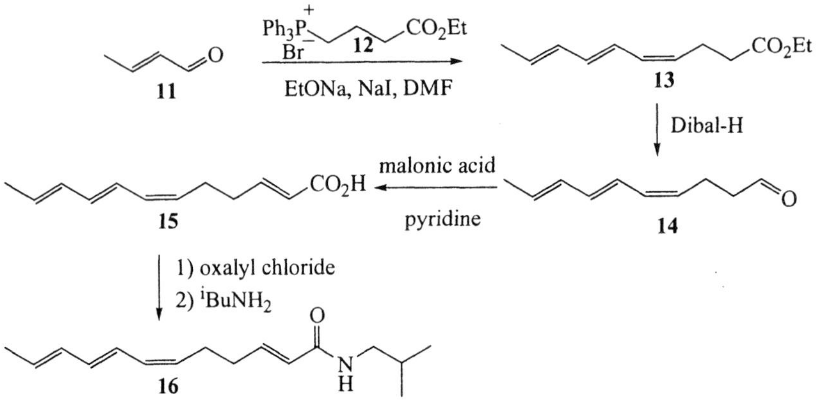

Scheme 1: Synthesis of α-sanshool

As a result of our observations we propose the model in Figure 4 as a good representation of the features require for sanshool and bungeanool-type chemicals to be pungent.

Figure 4 shows that the N-isobutylcarboxamide and the (CH=Z=CH-CH2-CH2-CH=E=CH) motif are required for pungency. However, some optional features enhance the pungent character of the molecule (hydroxyl group on N-alkyl group, unsaturation, longer chain). We noted that pungency was only noticeable when molecules had the required features plus two optional features.

To test our model we designed two new molecules **28** and **29** that follow the rules shown in Figure 4.

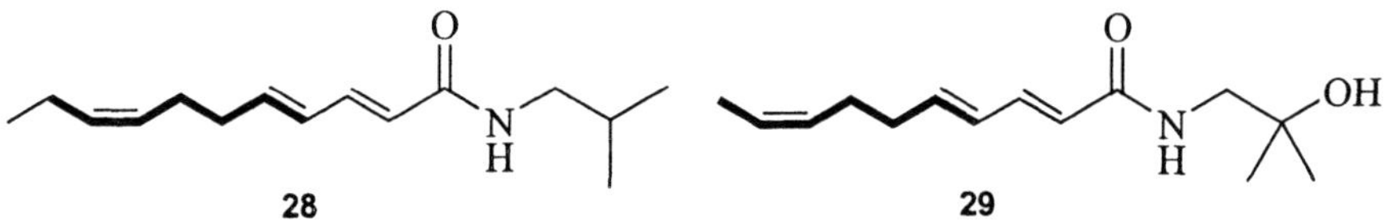

28 **29**

These molecules were very interesting because we expected that the absence of a triene system would stabilize the molecule, these structures had never been reported before, and their fatty chain could easily be made from commercially-available aldehydes **29** and **31** (Scheme 3). Amide **28** was of particular interest since we could synthesize it in one step by modification of a published method (Scheme 3) (*11*).

This experiment allowed us to validate our model since **28** and **29** exhibited a pungency typical of that of the sanshool and bungeanool families. Neat amides **28** and **29** showed enhanced stability properties probably because they could be obtained as white powders instead of thick oils. The crystalline matrix may slow down the polymerization reaction. While **29** was more pungent, **28** was easier to synthesize thanks to a one-pot reaction.

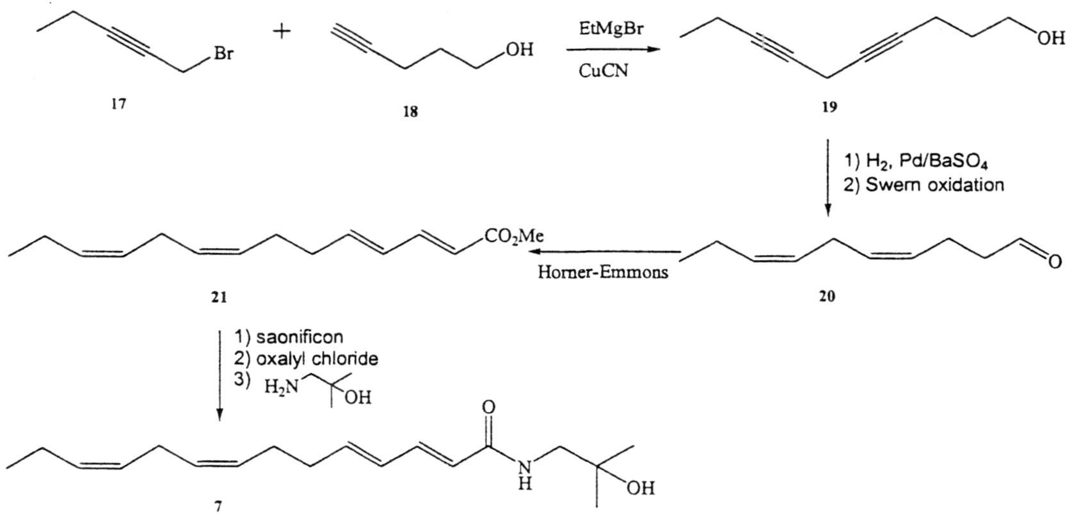

Scheme 2: Synthesis of bungeanool

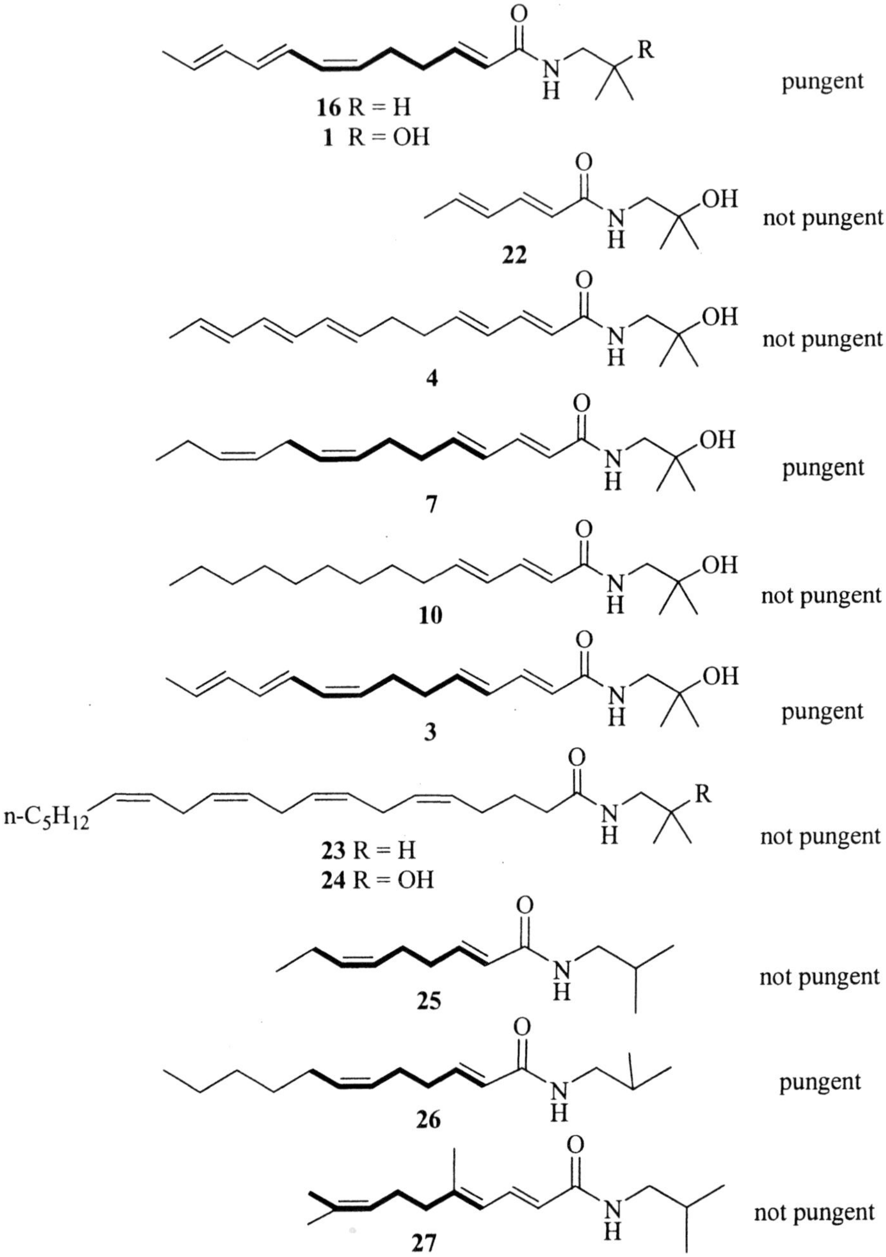

Figure 3 Activity of some sanshool and bungeanool derivatives

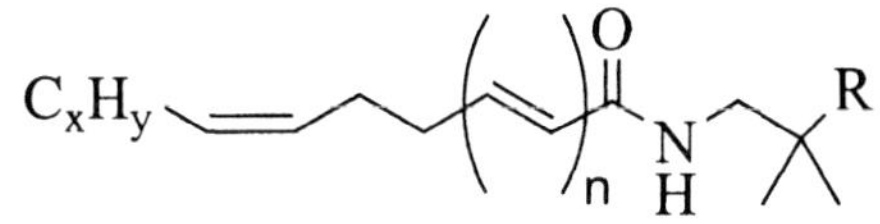

Minimal Structure: R=H, n=1, x=1

Optional Features: R=OH, n=2, x > 2

Noticeable Pungency = Minimal Structure + two Optional Features

Figure 4 Proposed structural requirements

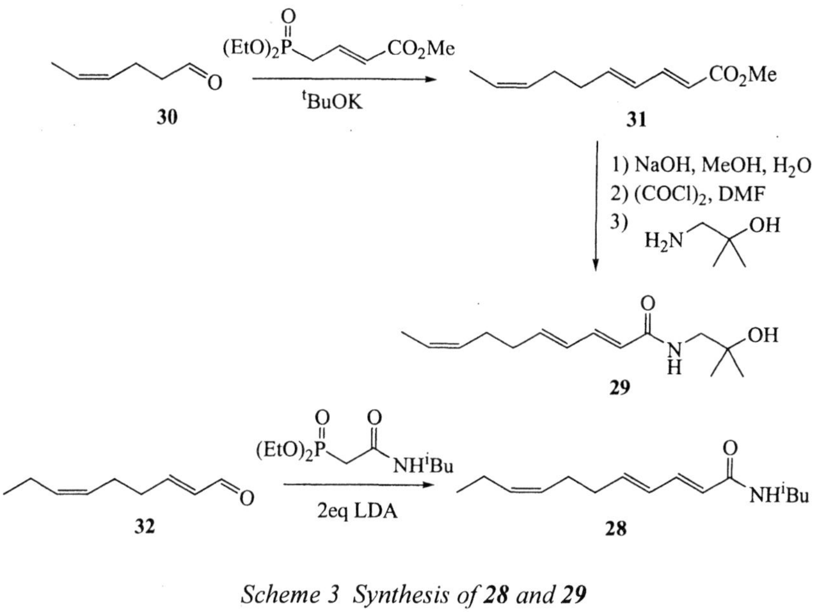

*Scheme 3 Synthesis of **28** and **29***

Application of the SAR Study of the Fatty Chain

It appeared to us that the only way to stabilize these molecules was to eliminate all unsaturation at the end of the chain. This was however impossible since we showed that at least one *cis*-double bond is needed at the end of the chain. We came up with the idea that the *cis*-double bond may be replaced by a phenyl group. In fact we realized that there may be three different ways that the chain could wrap up so that it would sterically look like a phenyl group (Figure 5).

To test this idea a series of phenyl-containing amides was prepared, a few examples are shown in Figure 6.

All the chemicals made in this series were stable at ambient temperature but only **35** showed a noticeable pungency. As far as we know, this simple cinnamamide had never been described as being pungent. Unfortunately, 35 is very weak and does not have the tingling character of the sanshool and bungeanool families.

SAR Study of the N-Alkyl Group

Although **35** was not a viable solution to our stability problem, it gave us a very accessible skeleton that we could use to vary the N-alkyl group. A series of cinnamamides was synthesized with the N-alkyl groups shown in Figure 7.

None of the new cinnamamides was pungent, which indicated that the isobutyl group was indeed required for pungency.

Conclusion

We have shown that a model could be built to predict whether a polyenamide of the sanshool or bungeanool type was pungent or not. We were able to validate the model by synthesizing unknown structures that fit the model and were found to be pungent. We were able to adapt the model to create a phenyl containing amide that had a pungent character. The activity of cinnamamide **35** suggests that sanshools and bungeanools must be activate their corresponding heat and tactile receptors as conformers were the fatty chain is wrapped up to activate. This theory will be difficult to prove until we have more information about the receptor, with which these chemicals are binding.

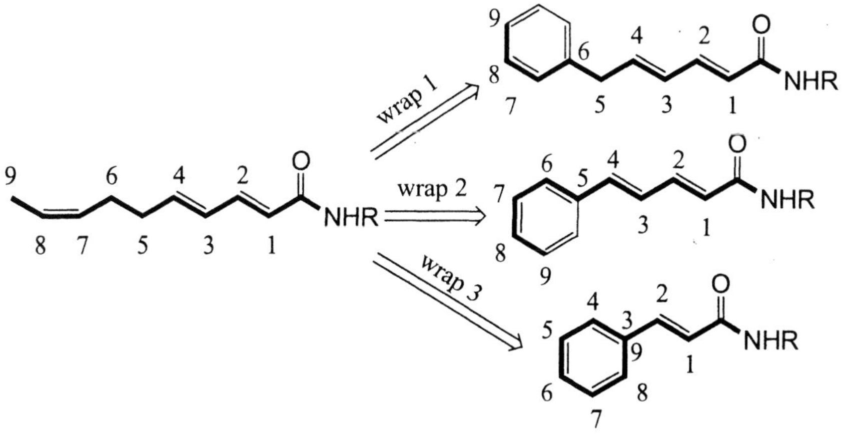

Figure 5 Wrapping of the fatty chain of a minimal pungent structure

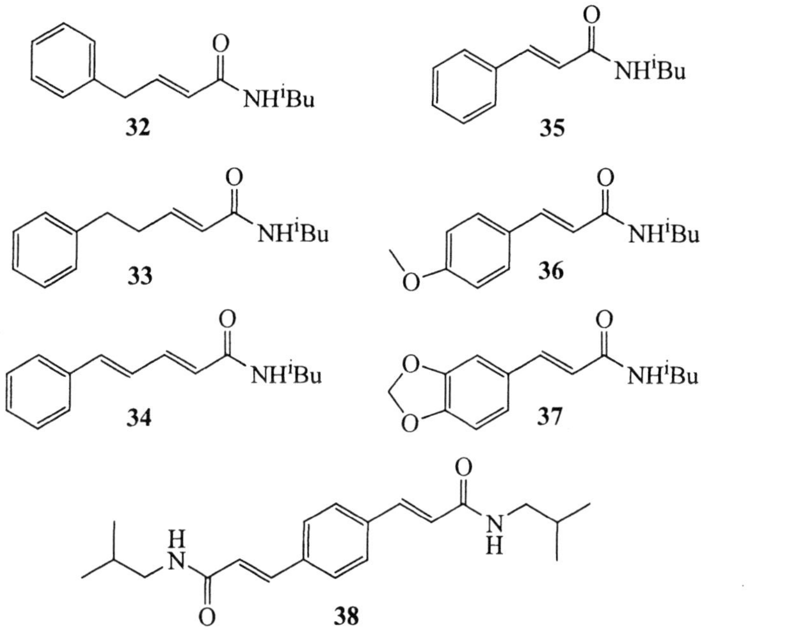

Figure 6 Some phenyl containing amides that fit the "wrapped" model

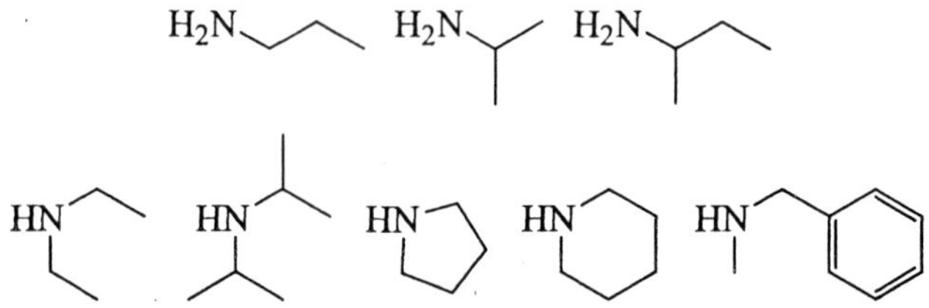

Figure 7 Various amines used to make cinnamamides

References

1. Bryant, B; Mezine, I. *Brain Research* **1999**, *842*, 452-460.
2. Sonnet, P. E. *J. Org. Chem.* **1969**, *34(4)*, 1147-1149.
3. Mizutani, K.; Fukunaga, Y.; Tanaka, O.; Takasugi, N.; Saruwatari, Y.-I.; Fuwa, T.; Yamauchi, T.; Wang, J.; Jia, M.-R.; Li, F.-Y.; Ling, Y.-K. *Chem. Pharm. Bull.* **1988**, *36(7)*, 2362-2365.
4. Crombie, L. *J. Chem. Soc.* **1955**, 995.
5. Crombie, L.; Tayler, J. L. *J. Chem. Soc.* **1957**, 2760.
6. Jacobson, M. *J. Org. Chem.* **1967**, *32*, 1646.
7. Xiong, Q; Shi, D.; Yamamoto H.; Mizuno, M. *Phytochemistry* **1997**, *46(6)*, 1123-1126.
8. Crombie, L. *J. Chem. Soc.* **1952**, 4338-4346.
9. Chen, I.-S.; Chen, T.-L; Lin, W.-Y.; Tsai, I.-L.; Chen Y.-C. *Phytochemistry* **1999**, *52*, 357-360
10. van der Linde, L. M.; van Lier, F. P.; van der Weerdt, A. J. A. European patent 0173395 A1, **1985**.
11. Crombie, L., Fisher, D. *Tetrahedron Lett.* **1985**, *26(20)*, 2477-2480.
12. Ward, J. P.; Van Dorp, D. A. *Recl. Trav. Chim. Pays-Bas* **1969**, *88*, 177.

Chapter 10

Structural Requirements for the Cooling Activity of Cyclic α-Keto Enamines

Thomas Hofmann[1], Tomislav Soldo[2], and Harald Ottinger[2]

[1]Institut für Lebensmittelchemie, Corrensstrasse 45, Westfälische Wilhelms-Universität Münster, D–48149 Münster, Germany
[2]Deutsche Forschungsanstalt für Lebensmittelchemie, Lichtenberstrasse 4, D–85748 Garching, Germany

3-Methyl- and 5-methyl-2-(1-pyrrolidinyl)-2-cyclopenten-1-one were recently identified as cooling compounds formed by Maillard reaction in heated glucose/L-proline mixtures, as well as in roasted malt. To gain more insights into the molecular requirements of this cooling effect, a range of cyclic keto enamine derivatives were synthesized, and their physiological cooling activities were evaluated. Any modification of the amino moiety, the carbocyclic ring size, or the alkyl substitution led to an increase of the cooling threshold. Insertion of an oxygen atom into the carbocycle, however, increased the cooling activity, e.g. the threshold of the corresponding3(2*H*)-furanone was 16-fold below the threshold concentration found for the 2-cyclopenten-1-one. Incorporation of an oxygen into the 5-position of the cyclopentenone resulted in a even more drastic effect, e.g. the 2(5*H*)-furanone exhibited the strongest cooling effect at the low threshold of 0.02-0.06 mmol/L. In contrast to the minty smelling (–)-menthol, these keto enamines are odorless, but impart a cooling sensation to the oral cavity and to the skin.

The mint-like smelling monoterpene (–)-menthol (**1** in Figure 1) is well-known to impart a "cooling" effect to the skin and the oral cavity and has long been added to food products such as chewing gums or mint sweets, as well as to cigarettes, in order to give a sensation of "freshness" during consumption. Furthermore, such cooling agents are used in cosmetic products to satisfy the demand for creating a feeling of "freshness" and "coolness" on the skin.

The sensation of cold on the skin and mucosal surfaces is neither caused by evaporation of the cooling agent, nor by vasodilation, that means there is not an actual temperature change inside the mouth, but is due to a specific action on sensory nerve endings (*1*). Calcium-imaging and patch-clamp studies have shown that cold promotes calcium influx, possibely through the direct opening of calcium-permeable ion channels (*2,3*). However, several other mechanisms have been proposed to explain cold-evoked membrane depolarization, including inhibition of background K^+-channels (*4*), activation of Na^+-selective degenerin channels (*5*), inhibition of Na^+/K^+-ATPases (*6*), and differential effects of cold on voltage-gated Na^+- and K^+-conductances (*7*). Subsequent studies suggested that (-)-menthol depolarizes sensory neurons by inhibiting Ca^{2+}-dependent K^+-channels (*8*), or by directly activating calcium-permeable channels (*2,9*).

The molecular biology of thermosensation has experienced a renaissance over the past few years, following the cloning and characterization of two ion channels that function as the principal heat sensors on mammalian somatosensory neurons. Besides the vanilloïd receptor channel subtype 1 (VR1), which is activated by moderate temperatures above 43 °C and by vanilloid compounds including capsaicin, the 'hot' component of chilli peppers, and the vanilloid-receptor-like channel type 1 (VRL-1) responding to temperatures above 50 °C, very recent investigations succeeded to identify a cDNA encoding a receptor that responded to the cold "mimic" menthol (*10*). In addition, this menthol-activated ion channel, which the authors named CMR1 (cold–menthol receptor type 1), was also activated by cold temperatures in the range of 8–30 °C. For the first time, these data provide a molecular explanation that cooling compounds and cold are detected by the same molecular entity on primary afferent neurons of the somatosensory system. Beginning with a bioinformatics approach new members of the TRP family were isolated, including TRPM8 (*11*). *In situ* hybridization experiments and functional studies then showed that this channel, which seem to be identical to CMR1, is a cold receptor expressed in a subset of somatosensory neurons that detect temperature and noxious stimuli.

To gain more detailed insights into the molecular requirements of chemicals interfering with these cold receptors, systematic sensory studies on more than 1200 synthetic cooling compounds were performed between 1971 and 1976 aimed at correlating their structure and their physiological cooling activity (*1*). Today various compounds are known to impart a "cooling" effect. This physiological effect seems to be not restricted to a distinct class of chemicals, but spans a range of different chemical classes, such as, e.g. from menthyl esters (e.g. **2**), carboxamides (e.g. **3**), menthane glycerol ketals (e.g. **4**), alkyl-substituted ureas (e.g. **5**), as well as 2,3-dihydro-1*H*-pyrimidine-2-ones (**6** in Figure 1) (*1,12*). Although most of these compounds exhibited cooling activity, none of these have yet been identified in nature.

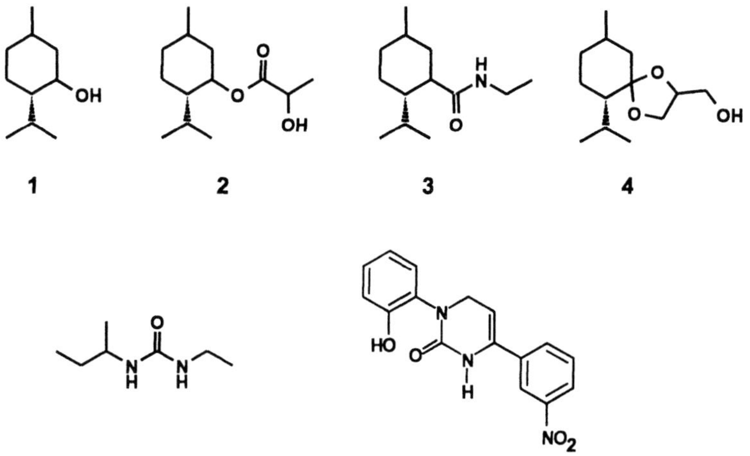

1 **2** **3** **4** **5** **6**

Figure 1. Structural diversity of cooling compounds reported in the literature (1,12).

Recently, the application of the taste dilution analysis, a novel bioassay-guided screening procedure enabling the identification of taste-active compounds in foods (*13*), led to the discovery of cooling compounds in dark beer malt as well as in roasted glucose/L-proline mixtures (*14*). MS and NMR studies as well as ^{13}C labeling experiments on the structures of the compounds contributing the most to the cooling sensation, followed by synthesis led to their unequivocal identification as 3-methyl-2-(1-pyrrolidinyl)-2-cyclopenten-1-one and 5-methyl-2-(1-pyrrolidinyl)-2-cyclopenten-1-one (*14*). Because these compounds belong to a novel class of cooling agents, it is as yet not known which molecular features of these cyclic α-keto enamines are required to exhibit a strong physiological cooling activity. The purpose of the present investigation was, therefore, to perform structure/activity studies on a series of synthetic α-keto enamines structurally related to the 2-amino-2-cyclopenten-1-ones.

Experimental

Syntheses of cyclic keto enamines

Keto enamines **7-27** and **29-32** were synthesized by refluxing the corresponding cyclic enolone and the corresponding amino compound in ethanol in the presence of acetic acid as reported recently (*15*). After purification by column chromatography using Al_2O_3 (basic activity III-IV, Merck, Darmstadt, Germany) as the stationary phase and pentane/diethyl ether mixtures as the mobile phase, the target compounds were obtained as colorless oils in purities of more than 99% and characterized by GC/MS as well as ^{1}H and ^{13}C NMR spectroscopy. Keto enamine **28** was preparatively isolated from a heated mixture of glucose and L-proline.

Sensory Analyses

Prior to sensory analysis, the purity of the synthetic compounds was proven by GC/MS as well as ^{1}H-NMR spectroscopy. Oral cooling thresholds were determined in a triangle test using tap water as the solvent (*15*). The samples (5 mL) were presented in order of increasing concentrations (serial 1:1 dilutions) and the threshold values evaluated in three different sessions were averaged. The values between individuals and separate sessions differed not more than one dilution step.

For topical testing aqueous solutions (0.2 mL) of the coolant were applied to a circular area (about 10 cm^2) of the skin surface on the inside of a forearm, midway between the wrist and the elbow, and were rubbed for 30 s. In parallel, an aliquot (0.2 mL) of pure tap water was applied as the blank onto the skin of the other forearm. After 30 s, the skin was dried with a towel. A panel of five male and five female subjects was asked to identify the arm where a cooling sensation was detectable, and furthermore, to rank perceived cooling intensity on a scale from 0 (no effect) to 5 (very strong). The values evaluated in three different sessions at two days were averaged. The values between individuals and separate sessions differed not more than two scores.

Results and Discussion

The 2-amino-2-cyclopenten-1-ones **7** and **8** (Figure 2), recently identified as the most active cooling compounds in dark malt formed by Maillard reactions from glucose and L-proline, were found to be virtually odor- and taste-less, but impart a long-lasting sensation of "cooling" and "freshness" to the oral cavity (*14*). To gain insights into the molecular requirements of this class of chemicals for imparting a cooling sensation, a range of derivatives varying either in the amino moiety, the ring size, or the number and position of methyl groups as well as oxygen atoms were synthesized, purified and their physiological cooling activities were evaluated sensorially. The tongue was found to be very sensitive to the cooling sensation of **7** and **8** (*10*), being well in line with the observation that cooling compounds applied on the tongue appear to penetrate rapidly to the cold nerve receptors (*1*). For that reason it is believed that the oral activities are a reasonable close approximation to the intrinsic activities of cooling compounds. Determining the oral thresholds of the cooling effect, therefore, appeared to give the best measure of intrinsic cooling activity.

Influence of amino moiety on cooling activity

Substitution of the pyrrolidine ring in compounds **7** and **8** (Figure 2) with either piperidine, diethylamine, dibutylamine, or dihexylamine, respectively, was done by a hydroxy/amine exchange from 2-hydroxy-3-methyl-2-cyclopenten-1-one and the corresponding amine in the presence of acetic acid in refluxing ethanol. The reaction products, namely 3-methyl-2-(1-piperidinyl)-2-cyclopenten-1-one (**9**), 5-methyl-2-(1-piperidinyl)-2-cyclopenten-1-one (**10**), 3-methyl-2-diethylamino-2-cyclopenten-1-one (**11**), 5-methyl-2-diethylamino-2-cyclopenten-1-one (**12**), 5-methyl-2-dibutylamino-2-cyclopenten-1-one (**13**), and 5-methyl-2-dihexylamino-2-cyclopenten-1-one (**14**) were purified by column chromatography and characterized by mass spectrometry as well as by 1D- und 2D-NMR experiments, and the cooling thresholds were determined by a trained sensory panel using triangle tests and water as the solvent (Figure 2).

The sensory evaluation revealed that enlargement of the pyrrolidine ring by one methylene group led to a significant increase of the cooling threshold, e.g. the cooling activity of **10** was 3 times lower than that found for compound **8** (Figure 2). Opening of the pyrrolidine ring showed no drastic influence on the cooling threshold, e.g. **12** showed only a 2-fold higher threshold when compared to **8**. However, further increase of the aliphatic amino moiety by two methylene groups led to a drastic decrease of cooling activity, e.g. a 9 or 4 times lower cooling activity was determined for **13** when compared to **8** or **12**, respectively. Increasing the chain length of the amino moiety to six carbon atoms completely diminished the physiological cooling perception, e.g. **14** did not impart any cooling effect but produced a burning, tingling-like sensation in the oral cavity.

To study the role of oxygen atoms in the amino moiety, in addition, 3-methyl-2-(1-morpholino)-2-cyclopentene-1-one (**15**), 5-methyl-2-(2-methoxy-carbonyl-1-pyrrolidinyl)-2-cyclo-pentene-1-one (**16**), and 3-methyl-2-(2-carboxy-1-pyrrolidinyl)-2-cyclopentene-1-one (**17**) were synthesized, purified and sensorially evaluated (Figure 2). Incorporation of the oxygen atom into the piperidine moiety of **10** nearly suppressed any cooling effect, e.g. **15** was evaluated with a 29-fold higher threshold concentration as the piperidine derivative **10** (Figure 2). Also the presence of the ester function in **16** and, more pronounced the free carboxylic acid group in **17** drastically decreased the cooling activity of compound **7**.

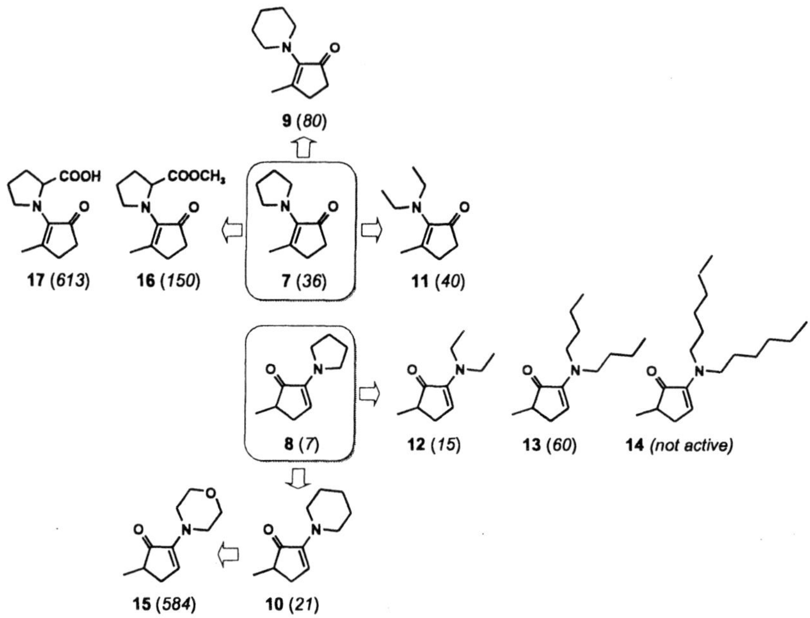

Figure 2. Cyclic α-keto enamines varying in the amino moiety; oral cooling threshold is given in brackets.

Influence of alkyl groups on cooling activity

To study how additional methyl groups or the substitution of a methyl group by an ethyl group are influencing the cooling activity of compound **8**, the derivatives **18-22** were synthesized (Figure 3). Sensorial evaluation showed that the absence of the methyl group in **18** as well as the insertion of an additional methyl group led to a significant increase of their cooling activity, e.g. the threshold of compounds **18** and **19** were by a factor of 7 above the threshold determined for compound **8**. The position of the second methyl groups influenced the cooling activities to a somewhat lower extend, e.g. **20** showed only by a 1.5 and 2 times lower cooling threshold than **21** and **19**, respectively

(Figure 3). Also the substitution of the methyl group by an ethyl group, lowered the cooling activity, e.g. **22** was 5 times less active when compared to compound **8** (Figure 3).

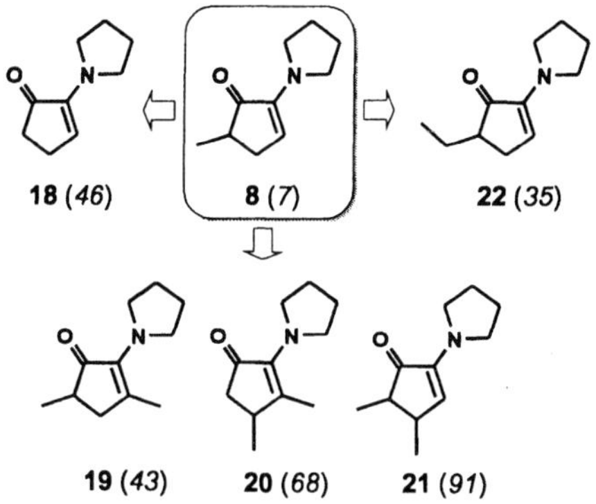

Figure 3. Cyclic α-keto enamines varying in the number and position of alkyl groups; oral cooling threshold is given in brackets.

Influence of position of the amino group and ring size on cooling activity

Changing the positions of the methyl group and the amino moiety in the molecule resulted in a total loss of cooling activity, e.g. the β-keto enamines **23** and **24** did not show any cooling effect (Figure 4).

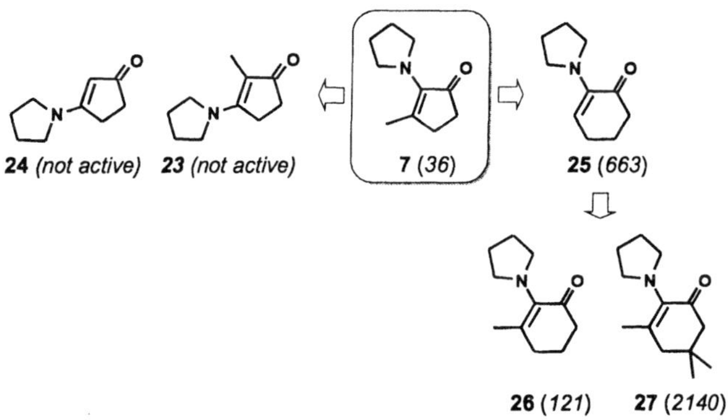

Figure 4. Cyclic β-keto enamines and α-keto enamines with six-membered carbocycle; oral cooling threshold is given in brackets.

To investigate the influence of the ring size of the carbocycle on the sensory attributes of the α-keto enamines, 2-(1-pyrrolidinyl)-2-cyclohexen-1-one (**25**), 3-methyl-2-(1-pyrrolidinyl)-2-cyclohexen-1-one (**26**), and 3,5,5-trimethyl-2-(1-pyrrolidinyl)-2-cyclohexen-1-one (**27**) were prepared and, after purification, the sensory data were compared with those found for the 2-amino-2-cyclopenten-1-one **7** (Figure 4). The results showed that ring enlargement of the 2-cyclopenten-1-one led to a significant decrease in cooling activity, e.g. **23** were evaluated with 3-fold higher cooling thresholds as the corresponding cyclopentenone **7** (Figure 4). Both, subtraction of the methyl group as well as as insertion of additional methyl groups, nearly quenched the cooling sensation.

Influence of number and position of oxygen atoms on cooling activity

In order to evaluate how the cooling activity is influenced by the presence of further oxygen atoms in the carbocycle, the carbaldehyde **28**, the 3(2*H*)-furanone **29** as well as the 2(5*H*)-furanone **30** were synthesized and sensorially evaluated (Figure 5). Oxidation of the methyl group of **7** to the carbaldehyde **28** diminished the cooling activity, but the substitution of the methylene group at C(4) in **7** resulted in a strong increase in cooling activity, e.g. the threshold of **29** is 16-fold below the threshold concentration determined for **7**, and reaches nearly the cooling threshold of 1.5 mg/L found for (–)-menthol (Figure 5). But in contrast to minty smelling (–)-menthol, compound **29** did not show any odor activity. In addition, the identification of **29** (4.4 μg/kg) in dark malt verified its natural occurrence in thermally processed foods, and demonstrated **29** as the most active, odorless cooling agent reported so far in nature.

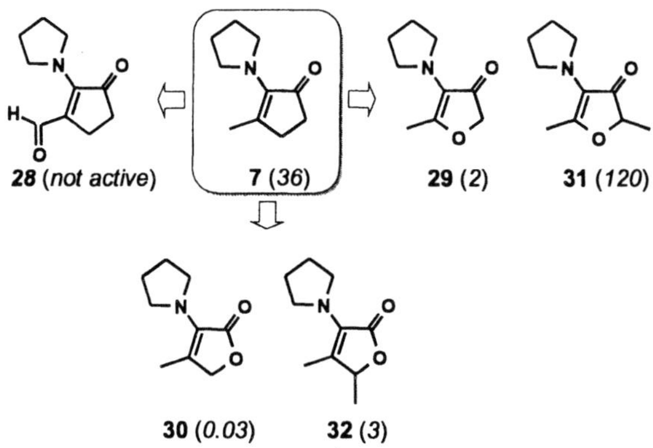

Figure 5. Cyclic α-keto enamines with additional oxygen in the carbocycle; oral cooling threshold is given in brackets.

Finally, the oxygen atom was shifted from the 4- into the 5-position of the cyclopentenone ring resulting in the γ-lactone **30** (Figure 5). This change in the molecule led to the most drastic increase in cooling activity, e.g. compound **30** exhibited a strong cooling effect at the low threshold concentration of 0.03 mg/L (Figure 5). Adding a second methyl group to these furanones led to a significant decrease of the cooling activity, e.g. a 60- or 100-times higher cooling threshold was found for **31** and **32** when compared to **30** and **32**, respectively (Figure 5).

Structural requirements for strong cooling activity of cyclic α-keto enamines

Taking all these data into consideration, it can be concluded that the cooling activity of compound **8** cannot be intensified any more by chemical modifications, whereas the physiological activity of the isomer **7** can be increased by the insertion of oxygen atoms, e.g. lactone **30** (X=-CH_2 and Y=O in Figure 6) was found as the most potent cooling compound among these α-keto enamines, e.g. the threshold concentration of **30** is lower by a factor of 35 or 906 than that found for (–)-menthol or **7** (X,Y=-CH_2 in Figure 6), respectively. The presence of two methyl groups is not favorable for the cooling activity independent from the position.

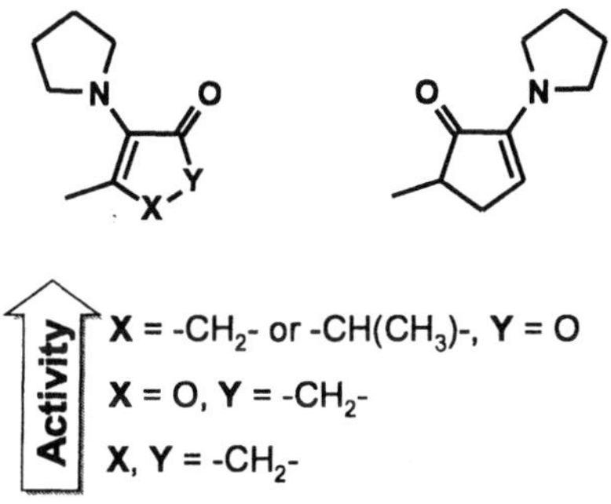

Figure 6. Structural requirements for cooling activity of cyclic α-keto enamines.

Cooling thresholds of selected α-keto enamines in milk chocolate

To gain first insights into the oral cooling activity of these keto enamines in foods, the detection thresholds of compounds **7**, **8**, **29**, **30** and **32** for imparting a cooling effect to milk chocolate was investigated. Serial dilutions of the keto enamines in milk chocolate were then presented to the sensory panel in order of increasing concentrations, and each dilution was sensorially assessed in a tringle test with to blanks (chocolate). Compound **7** was evaluated with the highest cooling threshold of 37.5 – 62.5 mg/100g chocolate, whereas **8** and **32** showed a 17 and 8 –fold lower cooling detection concentration (Table I). The lowest cooling thresholds in milk chocolate were found for compounds **29** and **30** with 0.8 – 1.5 and 0.25 – 0.50 mg/100g, respectively (Table I). The results obtained from this experiment show that the most active cooling compounds in water are also the most potent ones in milk chocolate, but the cooling thresholds are approximately 10-fold higher in milk chocolate compared to water.

Table I. Comparison of cooling thresholds of α-keto enamines in milk chocolate[a]

Compound	*Cooling threshold [mg/100g]*[a]
7	37.5 – 62.5
32	5.0 – 7.5
8	2.3 – 3.7
29	0.8 – 1.5
30	0.3 – 0.5

[a] Cooling threshold concentrations were determined in milk chocolate by means of a triangle test.

Topical cooling activity of cyclic α-keto enamines

In order to answer the question whether these novel α-keto enamines also impart a cooling sensation to the skin, topical tests were performed with compounds **8**, **29**, **30**, and **32** as well as with (–)-menthol as the control. Aliquots of aqueous solutions containing the cooling agent in increasing concentrations were, therefore, applied to a circular area of the skin surface on the inside of the forearm, and were rubbed for 30 s. In parallel, an aliquot of pure tap water was applied as the blank control onto the skin of the other forearm. The trained sensory panel was asked to rank the cooling intensity on a scale from 0 (no effect) to 5 (very strong). As given in Figure 7, all the compounds tested showed an significant cooling effect on the skin, however, they strongly differed in their cooling activities.

Compound **30** showed by far the highest cooling activity and was already topically detectable in a solution containing 0.00039% of the lactone (Figure 7). Increasing the concentration of **30** intensified the cooling sensation on the skin reaching the maximum score of 5 at a concentration of 0.00156%. In comparison, all the other compounds tested in that concentration did not impart any cooling effect. Further increase of the concentration revealed the compound **32** as the second effective coolant, which was topically detected when applied in concentrations of ≥ 0.0125% (Figure 7). In comparison, the 3(2*H*)-furanone **29** showed a significantly lower topical cooling activity, e.g. a significant cooling effect could be perceived in concentrations above 0.05% in water (Figure 7). Performing these topical experiments with (–)-menthol showed significantly lower effectivities for the terpene when compared to **29**, **30**, or **32**, e.g. the (–)-menthol imparted a cooling sensation in concentrations of ≥ 0.2%, which is above the topical detection threshold found for compound **30** by a factor of 512. Taking these all these data into account, it might be concluded that, the lactones **30** and **32** in particular, showed, besides their strong oral cooling activities, also significantly higher topical cooling activities as well-known cooling agents such as, e.g. (–)-menthol.

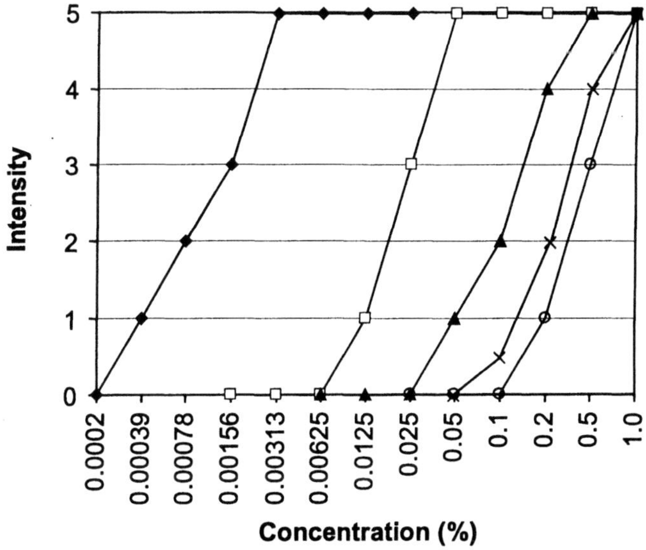

Figure 7. Topical testing of 30 (—◆—), 32 (—□—), 29 (—▲—), 8 (—O—), and (–)-menthol(—X—) on the inside of the forearm.

Conclusions

The data obtained for these cyclic α-keto enamine derivatives clearly show that the "cooling" activity as well as the odor quality change with variations in the amino moiety as well as in the carbocyclic ring, thereby illustrating that cooling and odor thresholds cannot be predicted from chemical structures, but have to be investigated based on systematic sensory studies with pure reference compounds. These investigations indicated that candidates of these cyclic α-keto enamine derivatives, in particular, **29**, **30**, **7** and **8**, might be used to evoke certain cooling effects during consumption of non-mint food compositions such as, drinking water, confectionary products, malted and citrus beverages and fruity or browned flavours. In additon, these compounds might be used as active cosmetic ingredients in hair shampoos, shower gels, roll-on deodorants, refreshing body lotions, as well as in oral and dental care products, providing a pleasant "freshness" and a long-lasting "cooling" effect on the skin, without the need for employing alcohol.

References

(1) Watson, H.R.; Hems, R.; Rowsell, D.G.; Spring, D.J. *J. Soc. Cosmet. Chem.* **1978**, *29*, 185-200.

(2) Reid, G.; Flonta, M.L. *Nature* **2001**, *413*, 480.

(3) Suto, K.; Gotoh, H. *Neuroscience* **1999**, *92*, 1131-1135.

(4) Reid, G.; Flonta, M.L. *Neurosci. Lett.* **2001**, *297*, 171-174.

(5) Akswith, C.C.; Benson, C.J., Welsh, M.J., Snyder, P.M. *Proc. Natl. Acad. Sci. USA* **2001**, *98*, 6459-6463.

(6) Pierau, F.K.; Torrey, P., Carpenter, D.O. *Brain Res.* **1974**, *73*, 156-160.

(7) Braun, H.A., Bade, H.; Hensel, H. *Pflugers Arch.* **1980**, *386*, 1-9.

(8) Swandulla, D.; Carbone, E.; Schafer, K.; Lux, H.D. *Pflugers Arch.* **1987**, *409*, 52-59.

(9) Okazawa, M.; Terauchi, T.; Shiraki, T.; Matsumura, K.; Kobayashi, S. *Neuroreport* **2000**, *11*, 2151-2155.

(10) McKerny, D.D.; Neuhausser, W.M.; Julius, D. *Nature* **2002**, *416*, 52-58.

(11) Viana, F.; De la Pena, E.; Belmonte, C. *Nature Neuroscience* **2002**, *5(3)*, 254-260.

(12) Wei, E.T.; Seid, D. *J. Pharm. Pharmacol.* **1983**, *35*, 110-112.

(13) Frank, O.; Ottinger, H.; Hofmann T. *J. Agric. Food Chem.* **2001**, *49*, 231-238.

(14) Ottinger, H.; Bareth, A; Hofmann, T. *J. Agric. Food Chem.* **2001**, *49*, 1336-1344.

(15) Ottinger, H.; Soldo, T.; Hofmann, T. *J. Agric Food Chem.* **2001**, *49*, 5383-5390.

Chapter 11

Stability of Cyclic α -Keto Enamines with Cooling Effect

Christoph Cerny, Fabien Robert, and Renaud Villard

Nestlé Research Center, Nestec Ltd., 1000 Lausanne 26, Switzerland

The stability of three recently discovered cooling compounds, 5-methyl-2-(1-pyrrolidinyl)-2-cyclopenten-1-one (5-MPC), 5-methyl-4-(1-pyrrolidinyl)-3(2*H*)-furanone (5-MPF) and 4-methyl-3-(1-pyrrolidinyl)-2(5*H*)-furanone (4-MPF), was studied. The compounds were dissolved in deuterated chloroform or deuterated methanol/water, and stored at 20°C and 37°C for up to 3 months. The stability was followed by ^{1}H-NMR and GC-MS. The 5-MPC showed poor stability and isomerised into 3-methyl-2-(1-pyrrolidinyl)-2-cyclopenten-1-one (3-MPC). On the other hand, 5-MPF and 4-MPF were reasonably stable under both lipophilic and hydrophilic conditions and more than 75% survived at 37°C. The ^{1}H-NMR and GC-MS data suggest that in all 3 compounds the protons at the cyclopentene and furanone rings are susceptible to hydrogen/deuterium exchange.

Recently, three new compounds eliciting a cooling sensation in the oral cavity and on the skin have been isolated from a Maillard reaction mixture by application of taste dilution analysis. The cooling compounds were identified in a thermally treated glucose/L-proline mixture as 3-methyl-2-(1-pryrrolidinyl)-2-cyclopenten-1-one (3-MPC), 5-methyl-2-(1-pyrrolidinyl)-2-cyclopenten-1-one (5-MPC), and 2,5-dimethyl-4-(1-pyrrolidinyl)-3(2*H*)-furanone (2,5-DMPF)(*1*). The chemical structures and the cooling thresholds in water are shown in Figure 1. Their presence was verified in dark malt by GC-MS and hence they can be considered as naturally occurring in food. They all have a cyclic α-keto enamine structure. A systematic study of the relationship between the structure and the physiological cooling property revealed 23 additional cyclic α-keto enamines with cooling properties (*2*). Two of them were described as odorless: 4-methyl-3-(1-pyrrolidinyl)-2(5*H*)-furanone (4-MPF) and 5-methyl-4-(1-pyrrolidinyl)-3(2*H*)-furanone (5-MPF). The latter could be identified in dark malt.

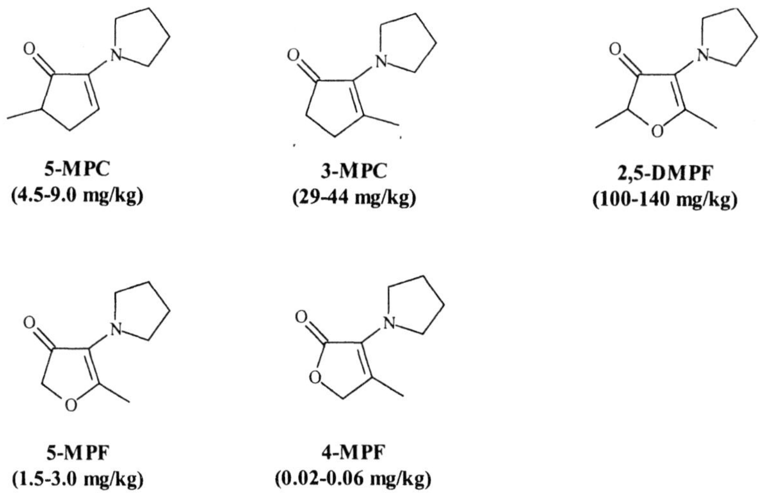

5-MPC (4.5-9.0 mg/kg)

3-MPC (29-44 mg/kg)

2,5-DMPF (100-140 mg/kg)

5-MPF (1.5-3.0 mg/kg)

4-MPF (0.02-0.06 mg/kg)

Figure 1. Cooling compounds with cyclic α-keto enamine structure; cooling thresholds from (2) are given in brackets

The goal of this work was to study the stability of three of these compounds with low cooling thresholds during storage at 20°C and 37°C. The selected compounds are 5-MPC, 5-MPF and 4-MPF.

Experimental

Materials

Pyrrolidine, chloroform-d (99.8% deuterium), methanol-d_4 (99.8% deuterium) and deuterium oxide (99.9% deuterium) were from Aldrich (Buchs, CH). Diethyl ether, hexane, methanol and toluene were of analytical grade and from Merck (Darmstadt, D). Aluminum oxide (B, Super I) was from Woelm (Eschwege, D).

The 5-MPC was synthesized from 2-hydroxy-3-methyl-2-cyclopenten-1-one and pyrrolidine (*3*) and 5-MPF from xylose and pyrrolidine (*4*). Both compounds were purified as described by Ottinger et al. (*2*).

Synthesis of 4-MPF

Fifty g 3-hydroxy-4-methyl-2(5*H*)-furanone (norsotolone) was synthesized according to Fleck et al. (*5*). A solution of norsotolone (40.0 g) and pyrrolidine (30.0 g) in toluene (500 ml) was refluxed for 60 min under nitrogen using a Dean-Stark system. The theoretical amount of water was obtained. After cooling to room temperature, the solvent was removed in vacuo. The resulting oil was distilled under vacuum (5-6 Pa, 126-127°C) and 41.3 g of product was obtained. The product contained 4-MPF and an unknown isomer. An aliquot of 35.0 g of the product was applied on a column (70 x 250 mm) filled with a slurry of aluminum oxide (1000 g, basic, activity III-IV) in hexane. Chromatography was performed using hexane (1000 ml; fraction A), hexane/diethyl ether (90/10, v/v; 1000 ml; fraction B), hexane/diethyl ether (80/20, v/v; 1000 ml; fraction C), hexane/diethyl ether (70/30, v/v; 1000 ml; fraction D), hexane/diethyl ether (60/40, v/v; 1000 ml; fraction E) and hexane/diethyl ether (50/50, v/v; 1000 ml; fraction F). Fractions C-D contained the target compound 4-MPF. It was freed from solvent under vacuo affording a colorless oil. The yield was 57%. The spectroscopic data (MS [EI], ^{1}H-NMR, ^{13}C-NMR) correspond to published data (*2*).

Spectroscopic data of the unknown 4-MPF isomer: MS (EI): 167 (15; $[M]^+$), 98 (100), 70 (15), 69 (15), 56 (47), 55 (83), 42 (10), 41 (40), 39 (33);

Storage Stability Test

Each of the 3 cooling compounds 5-MPC, 5-MPF and 4-MPF was dissolved in chloroform-d and in a mixture of methanol-d_4 and deuterium oxide (2/1, v/v). The concentrations were between 4 and 8%. The compounds were stored in NMR tubes at 20°C and 37°C. They were closed with Teflon caps to avoid solvent loss. The storage time was up to 108 days (chloroform-d) and 98 days (methanol-d_4/deuterium oxide).

Analysis

GC-MS analyses were performed on a HP 6890 GC coupled to a 5973 MSD (transfer line heated at 280°C). The temperature program was 20°C, 70°C/min, 60°C, 4°C/min, 240°C (34 min). The sample volume of 0.5 μl was injected splitless onto a HP-PONA column (50 m x 0.20 mm x 0.50 μm film thickness, Hewlett Packard) and a DB-WAX column (60 m x 0.25 mm x 0.50 μm film thickness, J&W). Mass spectra were obtained in the EI mode at 70 eV. Solutions in methanol-d_4/deuterium oxide were diluted (1:5, v/v) with methanol prior to analysis.

The samples for ^{1}H NMR spectroscopy were prepared in WILMAD 528-PP 5 mm Pyrex NMR tubes, using heavy solvent (0.7 mL). The NMR spectra were acquired on a Bruker AM-360 spectrometer equipped with a quadrinuclear 5 mm probe head, at 360.13 MHz. All shifts are cited in ppm. One-dimensional ^{1}H-NMR was acquired as described earlier using standard conditions (*6*). All chemical shifts were referenced to residual solvent as an internal standard. All spectra were interpreted using the MestRe-C 2.3 software.

Results and Discussion

5-MPC

Solutions of 5-MPC in chloroform-d and methanol-d_4/deuterium oxide, were stored for up to 108 days and then analyzed by GC-MS. Table I shows the composition of the stored solutions. After storage at 20°C more than 50% of the 5-MPC was still present after 3 months. Storage at 37°C led to both strong hydrolysis with formation of pyrrolidine and 2-hydroxy-3-methyl-2-cyclopenten-1-one (cyclotene) and isomerisation to 3-MPC which has also cooling properties, but a much lower cooling threshold of 29-44 mg/kg vs. 4.5-9.0 mg/kg (*2*).

All stored solutions showed an intense brown color. Cyclotene, 4-hydroxy-5-methyl-3(2*H*)-furanone (norfuraneol) and pyrrolidine are discussed as chromophore precursors by Frank and Hofmann (*7*). A similar mechanism might be responsible for the color formation from 5-MPC.

Table I. Composition of 5-MPC Solutions After 3 Months Storage

Cooling Compound	*20°C (L)*	*37°C (L)*	*20°C (H)*	*37°C (H)*
5-MPC	75.2 %	27.3 %	52.2 %	36.9 %
3-MPC	14.1 %	36.6 %	12.8 %	32.6 %
Pyrrolidine	1.5 %	4.1 %	6.4 %	5.9 %
2-Hydroxy-3-methyl-2-cyclopenten-1-one	4.9 %	19.1 %	16.5 %	9.7 %
Others	4.3 %	12.9 %	12.1 %	14.9 %

L: in chloroform-d; H: in methanol-d_4/deuterium oxide (2:1, v/v)

Values are percentage areas in the GC-MS chromatogram (TIC, solvent was cut off)

Figure 2 shows the mass spectra of 5-MPC before and after 98 days storage at 37°C in methanol-d_4/deuterium oxide. Spectra in the EI mode were obtained from GC-MS analysis on a HP PONA column. They suggest an exchange of 1-4 protons against deuterium during storage.

Spectrum B represents a cluster of several deuterated compounds with molecular ion peaks of *m/z* 166-169 in contrast to the non-deuterated compound with *m/z* 165. The ions at *m/z* 94 and 95 in the spectrum of the non-deuterated 5-MPC can be attributed to the (1-pyrrolidinyl)-ethendiyl fragment of the molecule. The corresponding signals in the deuterated compound *m/z* 94, 95 and 96 indicate a partial deuteration at the C-3 atom. The signal at *m/z* 108 in the non-deuterated 5-MPC can be related to the (1-pyrridinyl)-1-propendiyl fragment. the corresponding signals in the deuterated compound (*m/z* 108, 109, 110 and 111) indicate a mixture of non, mono, double and triple deuteration of this fragment. Finally, the signal at *m/z* 150 in the non-deuterated 5-MPC is likely the result of the abstraction of the methyl group at C-5. The corresponding peaks *m/z* 150, 151 and 152 in the spectrum of the deuterated molecule indicate the loss of *m/z* 15, 16 and 17 suggesting none, mono and double deuteration of the methyl group. This is in agreement with the observation that most of the protons in cyclic enolones are acidic and therefore exchangeable for deuterium.

These findings suggest that the 5-MPC in methanol-d_4/deuterium oxide stored at 37°C for 98 days is a mixture bearing 1-4 deuterium atoms. The position of the labeling is at the cyclopentene ring and the methyl group. On the other hand, deuteration at the pyrrolidinyl ring does not occur under these conditions.

The ^{1}H-NMR spectrum clearly confirms the presence of deuterium in the structure. In chloroform-d, the aliphatic protons of the five-member ring show an AMX system at 2.05, 2.35 and 2.71 ppm (cf. Figure 3A). The proton in the position relative to the carbonyl group is represented at 2.35 ppm by a ddq. The two geminal protons are represented by ddd at 2.05 and 2.71 ppm. The geminal constant is ^2J= 17.75 Hz.

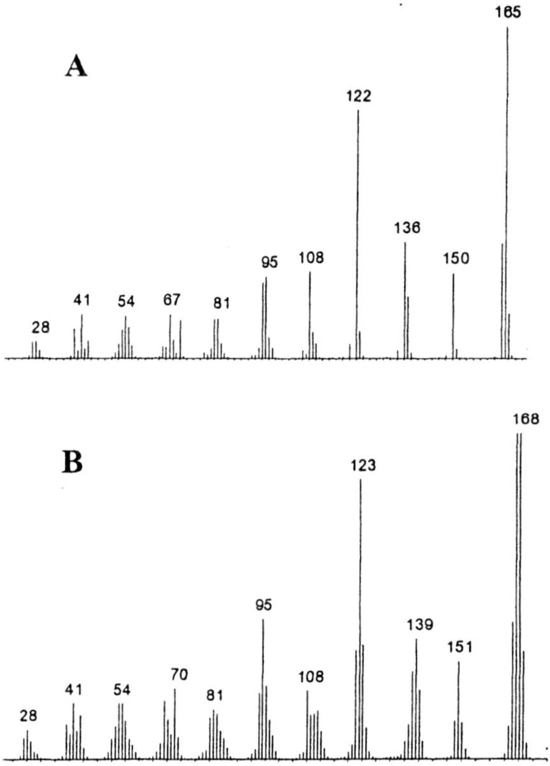

Figure 2. Mass spectra of A) 5-MPC and B) deuterated 5-MPC (after 98 days at 37°C in methanol-d_4/deuterium oxide)

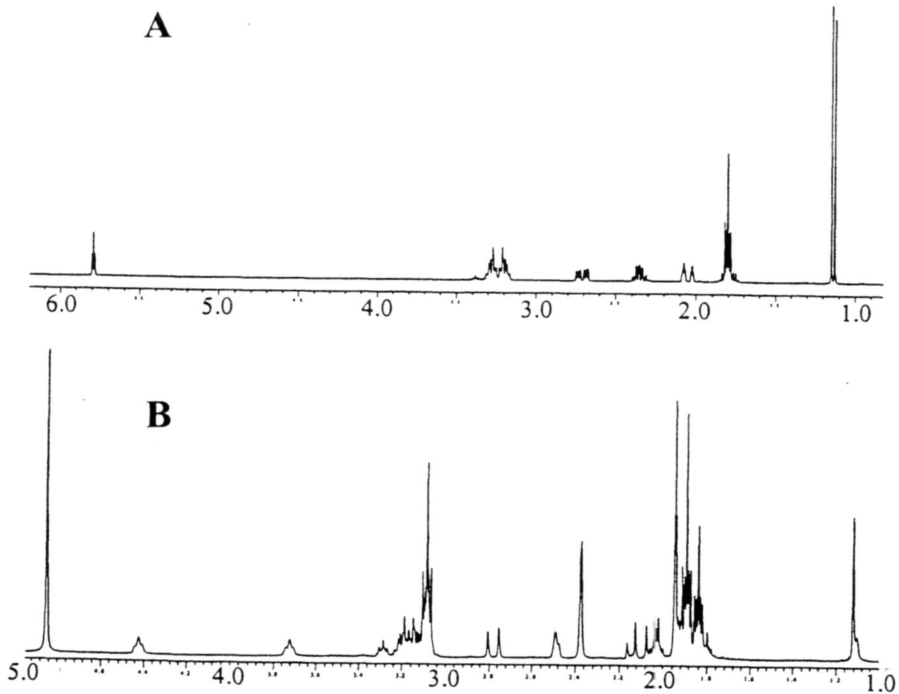

Figure 3. ^{1}H-NMR spectra of 5-MPC in A) chloroform-d and B) in methanol-d_4/deuterium oxide (after 13 days at 20°C)

In the ^{1}H-NMR spectrum of 5-MPC after 4 days in methanol-d_4/deuterium oxide (Figure 3B), the signal representing the proton in the position relative to the carbonyl has completely disappeared. The signal at 5.80 ppm (ethylenic proton) has also disappeared. This indicates that these two protons had been completely exchanged by deuterium after 4 days. This is confirmed by the shape of the two geminal proton signals; the AX system is now represented by two doublets with the ^2J constant coupling only. These observations confirm the hypothesis established *via* mass spectrometry and reveal the protons at carbon 3 and 5 as the most acidic ones in the molecule.

5-MPF

The composition of 5-MPF solutions stored in lipophilic and hydrophilic solvents is shown in Table II. The compound 5-MPF shows a reasonably good stability. Ninety-six percent of the 5-MPF were still intact after 98 days at 20°C in methanol-d_4/deuterium oxide. Storage at 37°C in both hydrophilic and lipophilic solvent caused a significant degradation, but still more than 80% of 5-MPF survived.

Table II. Composition of 5-MPF Solutions After 3 Months Storage

Cooling Compound	*37°C (L)*	*20°C (H)*	*37°C (H)*
5-MPF	85.0 %	96.0 %	81.5 %
Pyrrolidine	1.0 %	1.8 %	11.4 %
4-Hydroxy-5-methyl-3(2*H*)-furanone			0.3 %
Others	14.0 %	2.2 %	6.8 %

L: in deuterated chloroform; H: in methanol-d_4/deuterium oxide (2:1, v/v)

Values are percentage areas in the GC-MS chromatogram (TIC, solvent was cut off)

Pyrrolidine forms during hydrolysis and in particular is present in the sample which was stored at 37°C in methanol-d_4/deuterium oxide. Since pyrrolidine is an odor-active molecule with a relatively low aroma threshold of 20 mg/kg in water (*8*), it might have a negative impact on the aroma of stored food containing this cooling compound. The other hydrolysis product 4-hydroxy-5-methyl-3(2*H*)-furanone (norfuraneol) was present only at a low concentration and only in the hydrophilic solvent sample stored at 37°C. Possibly, it undergoes further degradation and contributes to the dark brown color formation in the stored solutions. Norfuraneol and pyrrolidine are mentioned as color precursors in the Maillard reaction during roasting of proline with xylose (*7*). The authors claim 5-MPF as an intermediate in the color formation.

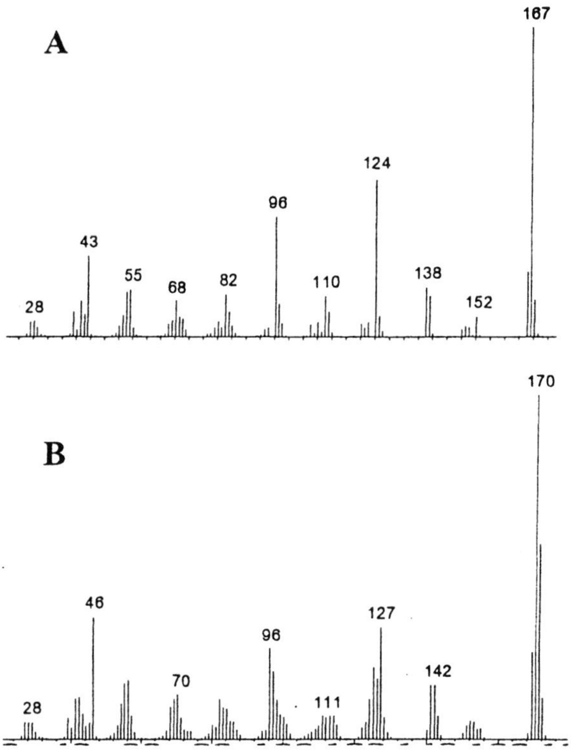

Figure 4. Mass spectra of A) 5-MPF and B) deuterated 5-MPF (after 98 days at 37°C in methanol-d_4/deuterium oxide)

Similar to 5-MPC, 5-MPF undergoes proton/deuterium exchange when stored in deuterium oxide containing solvents. Figure 4 shows the mass spectra of non-deuterated 5-MPF, as found after storage in deuterated chloroform, and the cluster of deuterated 5-MPF (GC-MS analysis of the 5-MPF solution in methanol-d_4/deuterium oxide after 98 days at 37°C). The molecular ion peak of 5-MPF is at *m/z* 167 (cf. Figure 4A). The signals in the spectra of the deuterated 5-MPF (cf. Figure 4B) at *m/z* 170 and 171 suggest a mixture of triple and quadruple deuterated compounds. Abstraction of the methyl group at the C-5 atom of 5-MPF in the mass spectrometer gives rise to the signal at *m/z* 152. The corresponding peaks *m/z* 153, 154 and 155 in the spectrum of the deuterated compound suggest the abstraction of *m/z* 16 and 17 indicating a mixture of compounds which are singly and doubly deuterated at the methyl group.

These results suggest a mix of triple and quadruple deuterated 5-MPF after storage in methanol-d_4/deuterium oxide (37°C, 98 days) with two deuterium at the C-5 atom and one or two deuterium atom at the methyl group.

Figure 5A shows the ^{1}H-NMR spectra of 5-MPF. The two protons in the α-position of the carbonyl group are represented at 4.35 ppm by a singlet. This signal disappeared in the spectrum of the 5-MPF which was stored in methanol-

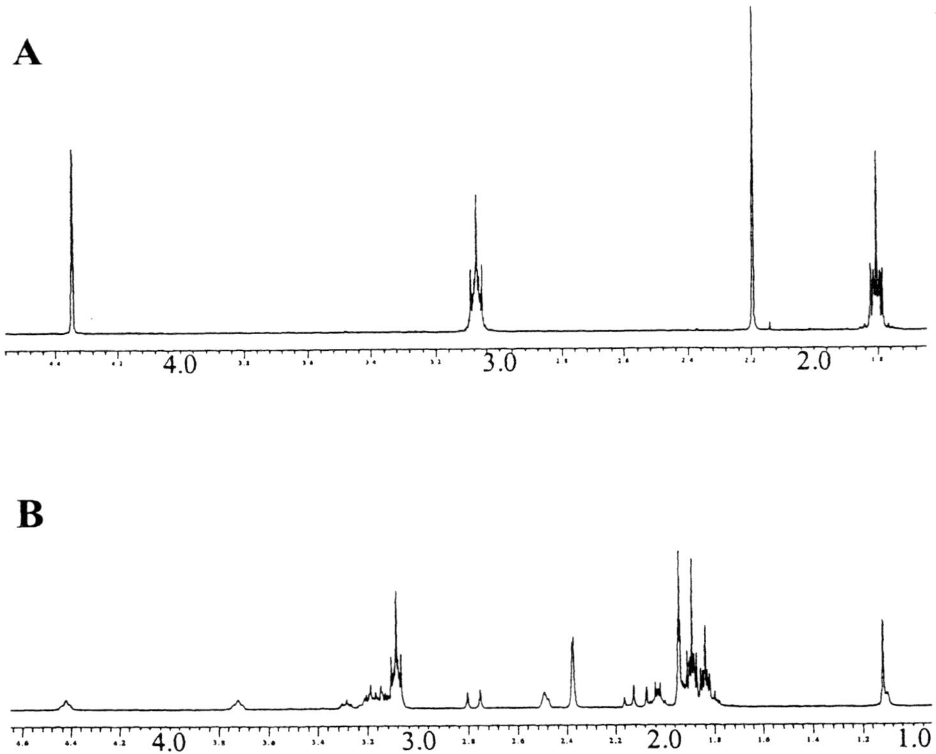

Figure 5 . [1]*H-NMR spectra of 5-MPF in A)* $CDCl_3$ *and B) in methanol-*d_4*/deuterium oxide (after 13 days at 20°C)*

d_4/deuterium oxide (20°C, 12 days) revealing that the two protons had been completely exchanged for deuterium and confirming the MS data.

4-MPF

The 4-MPF showed a good stability at 20 and 37°C in a lipophilic and at 20°C in a hydrophilic environment, with more than 90% of the compound still present at the end of the storage trial. At 37°C in methanol-d_4/deuterium oxide still 75% of the original compound were present. The main degradation mechanism is hydrolysis with pyrrolidine as the main decomposition product (cf. Table III)

A comparison of the spectra of 4-MPF before and after storage in methanol-d_4/deuterium oxide (Figure 6) shows a shift of the molecular ion signal of *m/z* 167 to higher values of *m/z* 168 and 169 indicating that an exchange of 1-2 protons against deuterium had happened. The same difference in mass could be observed when comparing the fragments *m/z* 122 (Figure 6A) and *m/z* 123 and 124 (Figure 6B). These masses correspond to the molecular ion after abstraction of formic acid. Loss of formic acid is typical for the fragmentation of isotetronic acids e.g. 3-hydroxy-4-methyl-2(5*H*)-furanone (*9*), which are similar in structure to 4-MPF.

Table III. Composition of 4-MPF Solutions After 3 Months Storage

Cooling Compound	*37°C (L)*	*20°C (H)*	*37°C (H)*
4-MPF	90.4 %	99.4 %	75.1 %
Pyrrolidine	2.7 %	0.6 %	14.3 %
3-Hydroxy-4-methyl-2(5*H*)-furanone	0.6 %		
Others	6.3 %		10.6 %

L: in chloroform-d; H: in methanol-d_4/deuterium oxide (2:1, v/v)

Values are percentage areas in the GC-MS chromatogram (TIC, solvent was cut off)

Supposedly the proton/deuterium exchange happened at the C-5 atom indicating a certain acidity of these protons, however probably much less than those of the 5-MPF protons. The low acidity and the low susceptibility of the lactonic carbonyl group to a nucleophile attack might explain the fact that the 4-MPF solutions remained colorless upon storage while the 5-MPF and 5-MPC solutions turned dark brown. ^{1}H-NMR data confirm the good stability of the compound since unlike with the other cooling compounds no significant changes were observed after 13 days storage in methanol-d_4/deuterium oxide (data not shown). Only after 98 days at 37°C was the exchange of the C-5 protons noticeable.

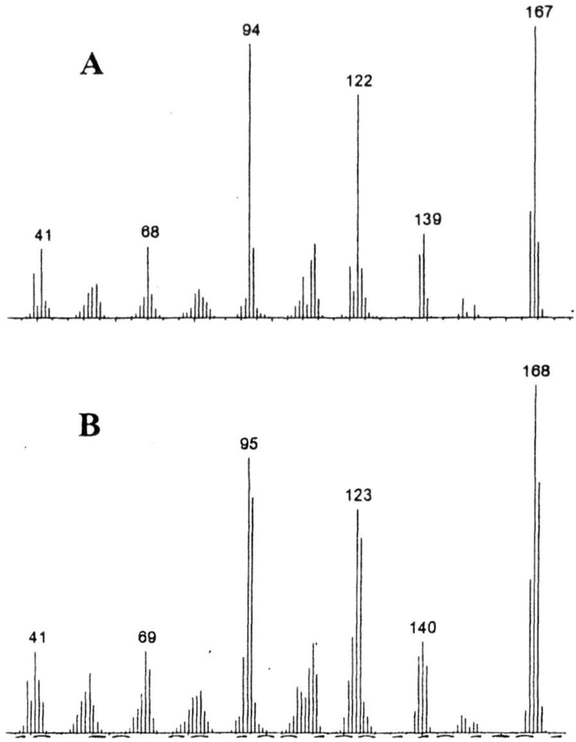

Figure 6. Mass spectra of A) 4-MPF and B) deuterated 4-MPF (after 98 days at 37°C in methanol-d_4/deuterium oxide)

Conclusion

The study revealed a poor stability for 5-MPC in lipophilic and hydrophilic solvents. Hydrolysis and isomerisation to 3-MPC with strong color development occurred during storage. Good stability was found for 5-MPF with more than 80% of the compound still present after 3 months at 37°C in both lipophilic and hydrophilic solvents. Nethertheless brown color formation was also observed. Similarly to 5-MPF, 4-MPF showed good stability during the experiment, however without coloration. All cooling compounds are susceptible to hydrolysis with liberation of pyrrolidine, especially in an aqueous environment at elevated temperatures.

Acknowledgments

We thank Drs. Francia Arce-Vera and Dieter Welti for NMR analysis and Dr. Elizabeth Prior for critically reading the manuscript.

References

1. Ottinger, H.; Bareth, A.; Hofmann, T. *J. Agric. Food Chem.* **2001,** *49,* 1336-1344.
2. Ottinger, H.; Soldo, T.; Hofmann, T. *J. Agric. Food Chem.* **2001,** *49,* 5383-5390.
3. GB Patent 1096427A, **1966**
4. Peer, H. G.; van den Ouweland, G. A. M.; De Groot, C. N. *Rec. Trav. Chim.* **1968,** *87,* 1011-1016.
5. Fleck, F.; Rossi, A.; Hinder, M.; Schinz, H. *Helv. Chim. Acta* **1959,** *33,* 130-139.
6. Lin, J.; Welti, D. H.; Arce Vera, F.; Fay, L. B.; Blank, I. *J. Agric. Food Chem.* **1999,** *47,* 2813-2821.
7. Frank, O.; Hofmann, T. *Helv. Chim. Acta* **2000,** *83,* 3246-3261.
8. Amoore, J. E.; Forrester, L. J.; Buttery, R. G. *J. Chem. Ecol.* **1975,** *1,* 299-310.
9. Bonini, C. C.; Iavarone, C.; Trogolo, C.; Poulton, G. A. *Org. Mass Spectrom.* **1980,** *15,* 516-519.

Taste-Active Peptides and Amino Acid Derivatives

Chapter 12

Savory Peptides Present in Moromi Obtained from Soy Sauce Fermentation of Yellow Soybean

Hanifah Nuryani Lioe[1], Anton Apriyantono[1,*], Dedi Fardiaz[1], Budiatman Satiawihardja[1], Jennifer M. Ames[2], and Elizabeth L. Inns[2]

[1]Department of Food Technology and Human Nutrition, Faculty of Agricultrual Technology, Bogor Agricultural University, P.O. Box 220, Bogor 16002, Indonesia
[2]School of Food Biosciences, The University of Reading, Whiteknights, Reading RG6 6AP, United Kingdom

The presence of savory peptides in moromi has been investigated. Moromi was prepared by fermenting yellow soybean using *Aspergillus oryzae* as the starter at the first step (mold fermentation) and 20 % brine solution at the next step (brine fermentation). The moromi was then ultrafiltered stepwise using membranes with MW cut-offs of 10,000, 3,000, and 500 Da, respectively. The fraction with MW < 500 Da was chromatographed using Sephadex G-25 SF to yield four fractions, 1-4. Analysis of soluble peptides, NaCl content, α-amino nitrogen, amino acid composition, peptide profile using CE coupled with DAD, taste profile and free glutamic acid content, were performed for each fraction. Fraction 2 contained a relatively high total glutamic acid content, but a relatively low free glutamic acid content and had the highest umami taste. This fraction also had more peptides containing non-aromatic amino acids than the other fractions. The peptides present in fraction 2 may play a role, at least in part, in its intense umami taste.

Introduction

It is known that many peptides have taste, i.e., bitter, sweet, sour, salty and umami. These peptides were isolated from cheese (*11, 12*), beef broth (*31*), hydrolyzed soy protein (*17*), miso (*27*), soy sauce (*16*), fish protein hydrolysate (*20*), as well as synthesized in the laboratory (*21, 32*).

Savory peptides had interested scientists attention since Yamasaki and Maekawa reported the isolation of a "delicious" peptide from beef broth in 1978. The peptide was an octapeptide having the primary structure Lys-Gly-Asp-Glu-Glu-Ser-Leu-Ala. Later, this peptide was called as the "beefy meaty peptide" (BMP) and the synthesized one had a meaty (*31*), umami, savory (28) taste. Now, BMP is called STEP (savory taste-enhancing peptide), i.e. a peptide that is able to enhance savory taste (*26*).

Soy sauce is one of the fermented soybean products (*13*) which is rich in flavor due to either volatile (*2, 24*) or non-volatile compounds (*3, 4, 17, 22*). Apriyantono et al. (*4*) investigated sensory and peptide characteristics of soy sauce and its fractions obtained by ultrafiltration. They reported that peptides with molecular weight of less than 500 Da may play a role, at least in part, in increasing the intensity of the umami taste. It is very likely that some peptides having an umami taste are present in soy sauce, but they have not been identified yet. Therefore, attempts have been made to obtain the fractions containing peptides that are responsible for the umami taste, by further fractionation of moromi fraction with molecular weight of less than 500 Da. These fractions were then characterized chemically and sensorially and their peptides profiles were determined.

Materials and Methods

Materials

Materials for moromi preparation, i.e., yellow soybean and salt were obtained from a local market in Bogor (Indonesia). Dry culture of *Aspergillus sojae* was obtained from Konno Moyashi Co. (Kobe/Osaka, Japan).

Moromi Preparation

Yellow soybean was boiled in water (1:3 w/v) for one hour, cooled and drained. Cooked soybean was inoculated with 0.5 % (w/w) *Aspergillus sojae* starter. Inoculated soybean was incubated at room temperature for three days to obtain koji. Koji was soaked in 20 % brine solution (1:3 w/v) and then incubated at room temperature for eight weeks. The resulting fermentated product was called moromi. The moromi was homogenized, put in cheesecloth and then pressed by hand. The suspension obtained was centrifuged at 14 000 rpm for 10 minutes at 4 °C. Once centrifugation finished, the fat layer at the top was discarded, moromi liquid was then collected. Such preparation was done in three replicates.

Ultrafiltration Process

Moromi liquid was submitted to ultrafiltration cell at temperature 2–4 °C under 2–3 bar N_2 pressure using 50 mL stirred cell ultrafiltration unit (Amicon, Inc., Beverly, MA). Stepwise ultrafiltration was carried out using a 0.45 μm porosity membrane followed by YM 10 (molecular weight cut-off/MWCO 10 000 Da), YM 3 (MWCO 3 000 Da) and YC 05 (MWCO 500 Da) membranes (Amicon, Inc., Beverly, MA) . The fraction with molecular weight of less than 500 Da (Fraction MW<500 Da) was collected and lyophilized. This step was done for each replicate of moromi preparation.

Fractionation by Gel Filtration Chromatography

The lyophilized fraction with molecular weight of less than 500 Da was chromatographed on a column (2.5 x 57 cm) packed with Sephadex G-25 *superfine*. Deionized water was used as an eluant. Sixty tubes of eluate each with a volume 7.5 mL were collected. UV absorbance of each tube was measured at 240 nm. The absorbance was plotted against fraction and number and the resulting graph showed four peaks corresponding to four fractions, i.e. Fraction 1, Fraction 2, Fraction 3 and Fraction 4. The fractions were lyophilized. This step was done for each replicate of moromi. Analysis of soluble peptides (Lowry method), NaCl content (Mohr method), α-amino nitrogen (*1*), amino acid composition (*15*), peptide profile using CE coupled with DAD (see below), taste profile (see below) and free glutamic acid content (9), were performed for each fraction.

CZE Profile

Capillary zone electrophoresis was performed on a Hewlett Packard CZE system controlled by HP ChemStation software. Peptides was separated using an uncoated fused silica column (i.d. 50 μm, length 48.5 cm, effective length 40 cm) with conditions as follows: voltage, 15 kV; injection volume, 50 nL; running time, 30 min; temperature, 25 °C; buffer, phosphate 100 mM pH 2.5;

detection at 190–600 nm (Photodiode Array Detector/PDAD). Sample preparation was done by dissolving each lyophilized fraction in deionized water at the following concentrations: Fraction MW<500 Da, 0.0270 g/mL; Fraction 1, 0.0131 g/mL; Fraction 2, 0.277 g/mL; Fraction 3, 0.0022 g/mL; and Fraction 4, 0.0008 g/mL. Solutions were 0.22μm filtered, prior to analysis.

Taste Intensity

Fourteen trained panelists performed taste intensity analysis for both umami and salty taste. A series of standards for sensory taste intensity was used (Table I). Each lyophilized fraction was dissolved in deionized water so that the concentration of soluble peptides present in each fraction was 10.0 mg/10 mL, except Fraction 2 where the soluble peptides concentration was 1.0 mg/10mL. This is due to the high content of salt present in the fraction (59.9 mg NaCl/10 mL at the level of soluble peptides concentration 1.0 mg/10 mL). Taste intensity analysis was also performed at the same salt concentration as that for Fraction 2, i.e., 59.9 mg/10 mL.

Table I. Standards Used for Sensory Taste Intensity

Description	*Score*	*Standard*
Umami	50	0.05 % (b/v) MSG solution
	80	0.10 % (b/v) MSG solution
	150	0.20 % (b/v) MSG solution
Salty	50	0.25 % (b/v) sodium salt solution
	80	0.50 % (b/v) sodium salt solution
	150	1.25 % (b/v) sodium salt solution

Results and Discussion

Fractionation by Gel Filtration Chromatography

Moromi obtained from soy sauce fermentation of yellow soybean was ultrafiltered stepwisely to obtain Fraction MW<500 Da. Fraction MW<500 Da was then separated into four fractions by gel filtration chromatography using Sephadex G-25 superfine column. The initial aim of the fractionation was to desalinate the samples, but since it resulted 4 fractions (Figure 1), they were analysed further. Separation on Sephadex G-25 superfine is based not only on molecular size, but also on interaction between aromatic groups of the components and the matrix gel (*14*).

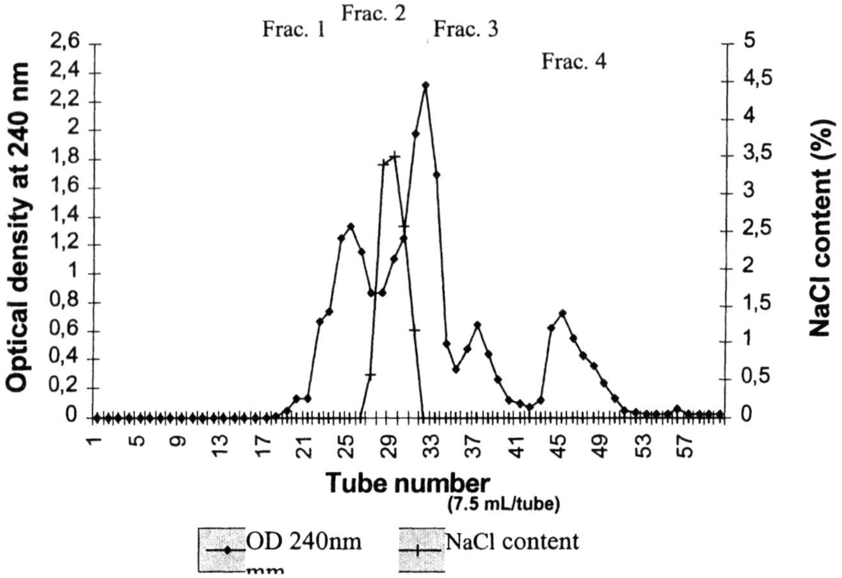

Figure 1. Gel filtration chromatogram of Fraction MW<500 Da obtained by moromi ultrafiltration. Fractionation was performed on Sephadex G-25 superfine column (57 cm x 2.5 cm); using deionized water as an eluant.

The fractionation of 1.000 gram Fraction MW<500 Da yielded 87±8 mg of Fraction 1, 781±16 mg of Fraction 2, 19±6 mg of Fraction 3 and 13±3 mg of Fraction 4. The fractions contained 10.7±1.9 mg, 11.1±2.0 mg, 7.6±1.4 mg, and 2.9±0.3 mg of total soluble peptides, respectively. Salt was present mainly in Fraction 2, at a concentration of 82.61±5.34 % (w/w). The results showed that the dry matter and peptides of Fraction MW<500 Da were mainly present in Fractions 1 and 2. Considering the separation mechanism on Sephadex G-25, these fractions are expected to have less aromatic compounds than the other fractions.

Amino Acid Composition and Peptide Profile

Amino acid composition of the fractions is shown in Table II. The amino acid present in Fraction MW<500 Da in the highest amount was glutamic acid. Glutamic acid, which was primarily present in Fractions 1 and 2. Table II also shows that aromatic amino acids, i.e. phenylalanine and tyrosine, were present in

Fractions 3 and 4 in much higher amounts than in the others. Tryptophan, another aromatic amino acid, was not analyzed, since it was destroyed during acid hydrolysis. Tryptophan and phenylalanine are hydrophobic amino acids that have intense bitter taste (*19*). Thus, it is more likely to find some peptides with umami taste or savory taste in Fractions 1 or 2.

Table II. Amino Acid Composition of Freeze Dried Fraction MW<500 Da and Its Four Fractions (Fraction 1, Fraction 2, Fraction 3, and Fraction 4)

Amino Acid	*Content, % (w/w)*				
	Fraction MW<500 Da	*Fraction 1*	*Fraction 2*	*Fraction 3*	*Fraction 4*
Asp	1.84 (18.4 mg)*	7.27 (6.3 mg)	0.86 (6.7 mg)	0.33 (0.1 mg)	1.20 (0.2 mg)
Glu	3.53 (35.3 mg)	15.48 (13.5 mg)	1.34 (10.5 mg)	0.71 (0.1 mg)	1.65 (0.2 mg)
Ser	0.54	1.12	0.33	0.25	0.94
His	0.36	0.78	0.19	0.23	0.38
Gly	0.62	1.28	0.35	0.59	1.02
Thr	0.59	1.42	0.34	0.40	0.62
Arg	0.10	0.50	0.05	0.11	0.71
Ala	0.79	1.78	0.45	0.54	1.02
Tyr	0.19	0.14	0.04	1.99	0.63
Met	0.18	0.06	0.16	0.06	-
Val	0.93	2.72	0.48	0.77	1.64
Phe	0.77	0.18	0.51	2.42	0.67
Ile	0.75	2.03	0.48	0.58	0.69
Leu	1.08	2.11	0.81	0.81	0.91
Lys	1.10	5.63	0.27	0.58	0.60
Total	13.37 (133.7 mg)	42.50 (37.0 mg)	6.66 (52.0 mg)	10.37 (2.0 mg)	12.68 (1.6 mg)

NOTE: values in parentheses are absolute weight based on the yield following fractionation of 1.000 g freeze dried Fraction MW<500 Da

Total glutamic acid of Fraction 1 was a bit higher than Fraction 2, but the free glutamic acid content of Fraction 2 was much lower than that of Fraction 1 (Table III). On the other hand, the amount of α-amino nitrogen present in Fraction 2 was about the same as in Fraction 1 (Table IV). This indicates that the glutamic acid present in Fraction 2 existed mostly as an amino acid residue within peptides.

Table III. Free Glutamic Acid Contents of the Four Fractions of Fraction MW<500 Da

Fraction	*Free glutamic acid content (%w/w)*
Fraction 1	19.52
Fraction 2	1.43
Fraction 3	Not detected
Fraction 4	Not detected

Table IV. α-Amino Nitrogen Contents of the Four Fractions of Fraction MW<500 Da

Sample	*α-Amino N Content (g NH_2/ 100 g peptide)*
Fraction 1	32.69±5.11
Fraction 2	31.04±6.73
Fraction 3	14.16±1.40
Fraction 4	7.10±0.52

Peptide profiles of the fractions are shown in Figure 2, and the spectral data obtained by the photodiode array detector (PDAD) over the range 190 – 600 nm are shown in Figures 3. The overall results showed that CZE effectively resolved peptides into several discrete peaks, except Fraction 4 which gave only one peak. The peptide profile of each fraction was different as shown by each electropherogram. The peptide peak could be recognized from the spectral data obtained by PDAD. The peptide bond has a high absorption at 214 nm (*6, 12, 14, 23*). Peptides containing aromatic amino acids also have high absorption over the 250 – 280 nm wavelength range (*5, 6, 8, 14*). On the other hand, peptides as well as free amino acids and organic acids have a carboxylic group which has high absorption in 185 – 208 nm wavelength range (*6, 8, 12, 23*). In fact, free amino acids and organic acids with molecular weights of less than 500 dalton have been reported previously in moromi (*3, 17*). Therefore, the presence of other compounds such as organic acids and amino acids, in this study cannot be ruled out.

Fraction MW<500 Da was resolved at least into nine peaks (Figure 2). Based on their spectral data they all contained peptides. The largest peak were 3, 5 and 6, of which Peak 5 had the highest intensity and did not contain aromatic amino acids as shown by its spectral data (Figure 3). On the other hand, both Peak 3 and 6 may contain aromatic amino acids. Peak 5 was the major peak of Fraction 2, Peak 6 was only detected in Fraction 3 and Peak 3 was only detected in Fraction 4.

As discussed above, Fraction 2 contained peptides containing glutamic acid residues while Peak 5 did not contain aromatic amino acids. Therefore, it is likely that Peak 5 is a peptide which contain glutamic acid residues and is expected to have a savory flavor.

Taste Intensity

Results of taste intensity analysis for umami and salty are shown in Figures 4 and 5. Taste intensity analysis for salty and umami was done at original salt concentration (each had different salt concentration), but the same concentration of soluble peptide content, i.e. 10 mg/10 mL, except for Fraction 2 which was 1 mg/10 mL (10 times more dilute than the other fractions). Fraction 1 had the highest score for umami taste, whereas Fraction 2, as expected, had the highest score for salty taste (Figure 4). Although Fraction 2 had been diluted 10 times as compared to the other fractions, Fraction 2 had relatively intense umami taste and its umami intensity may have higher intensity than that of Fraction 1 at the same soluble peptides concentration.

Because of the possible synergistic effect between salt and umami compounds, taste intensity analysis was also done at the same salt concentration, i.e. 59.9 mg/10 mL (salt was added to Fraction 1, 3 and 4), whereas the soluble peptide content was adjusted to 10 mg/10 mL except for Fraction 2 which was 1 mg/10 mL. The result (Figure 5) was similar to that for taste intensity analysis where only the soluble peptides content was adjusted, i.e., Fraction 1 had the highest umami intensity score followed by Fraction 2. Fraction 2 contained high amount of total glutamic acid, but low amount of free glutamic acid. In addition, peptide profile analysis using CE showed that Fraction 2 contained the main peptides present in Fraction MW<500 Da. Thus, Fraction 2 may contain savory peptides that are responsible for the intense umami taste of moromi.

Conclusions

The moromi fraction with MW<500 Da can be fractionated into four fractions, 1-4, by gel filtration on Sephadex G-25 superfine. Among these fractions, Fraction 2 had relatively high intensity of umami taste. It contained a

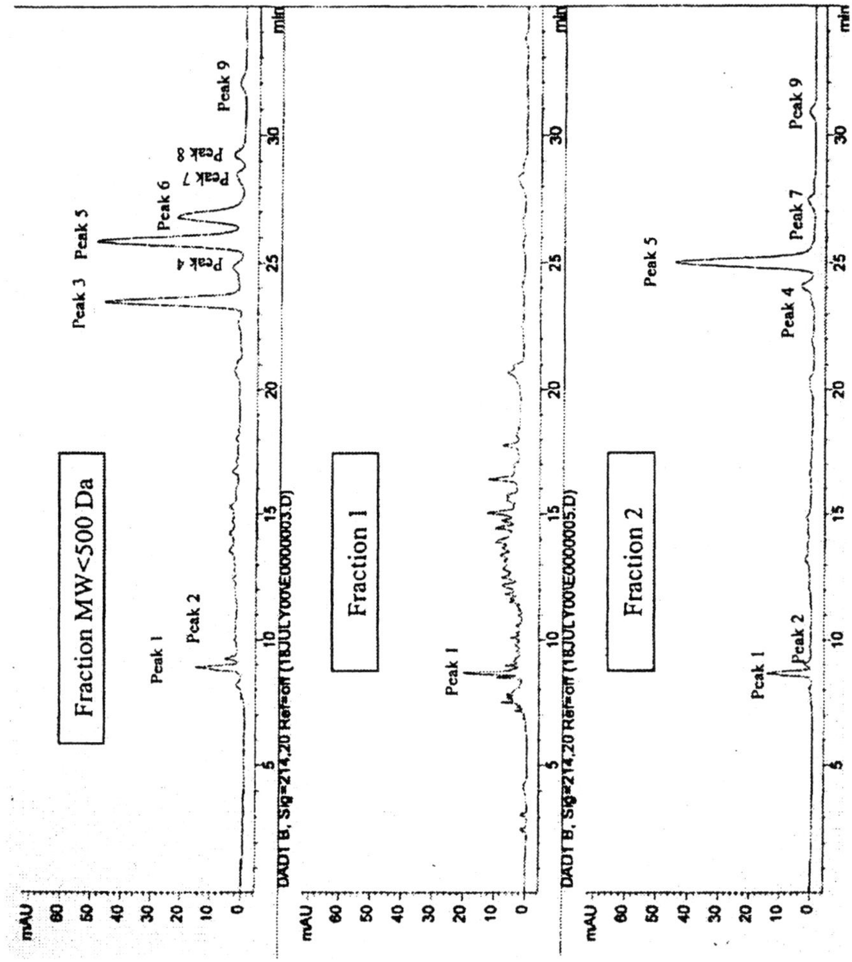
Fraction MW<500 Da
Fraction 1
Fraction 2
Peak 1
Peak 2
Peak 3
Peak 4
Peak 5
Peak 6
Peak 7
Peak 8
Peak 9
mAU
min
DAD1 B, Sig=214,20 Ref=off (18JULY00\E0000003.D)
DAD1 B, Sig=214,20 Ref=off (18JULY00\E0000005.D)

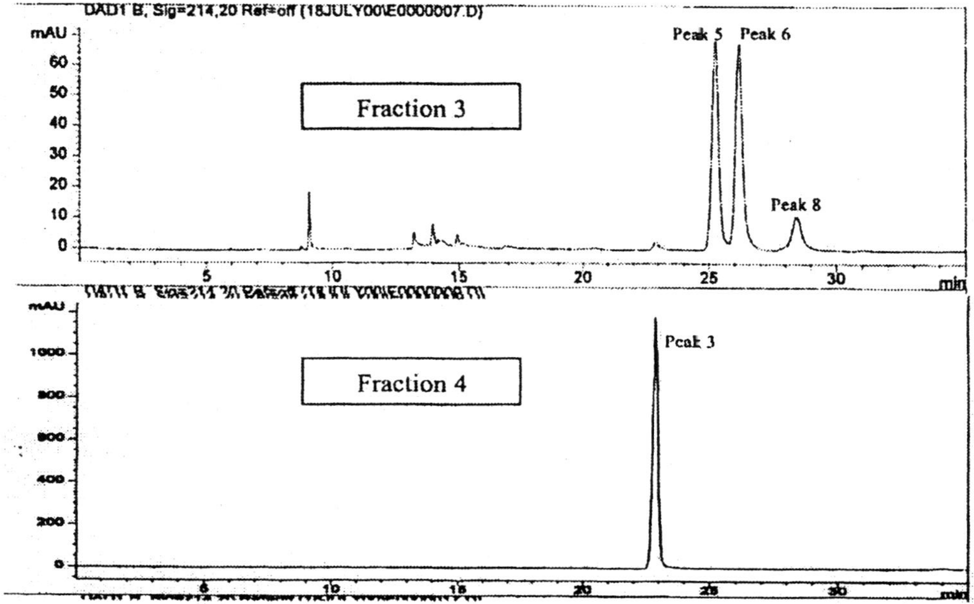

Figure 2. CZE Electropherograms of Fraction MW<500 Da and its four fractions (Fraction 1, Fraction 2, Fraction 3, and Fraction 4), employing detection at 214 nm; buffer phosphate 100 mM pH 2.5.

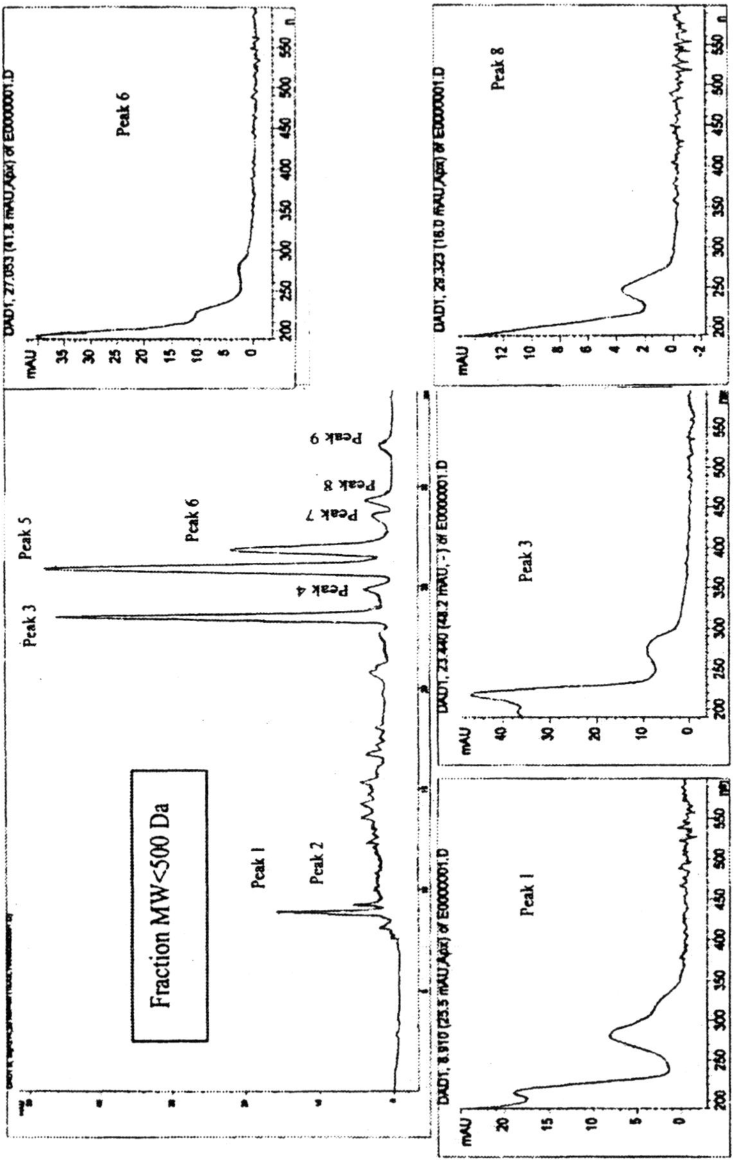

Fraction MW<500 Da
Peak 1
Peak 2
Peak 3
Peak 4
Peak 5
Peak 6
Peak 7
Peak 8
Peak 9
mAU

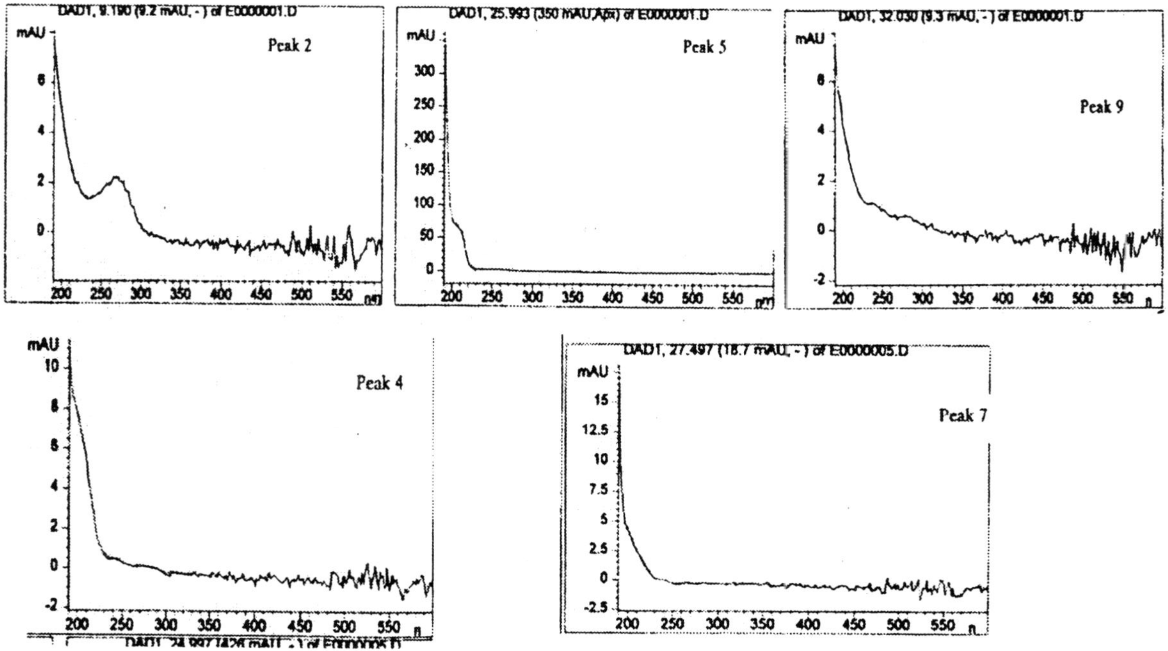

Figure 3. UV-Vis absorption spectra of peaks indicated in electropherogram of Fraction MW<500 Da.

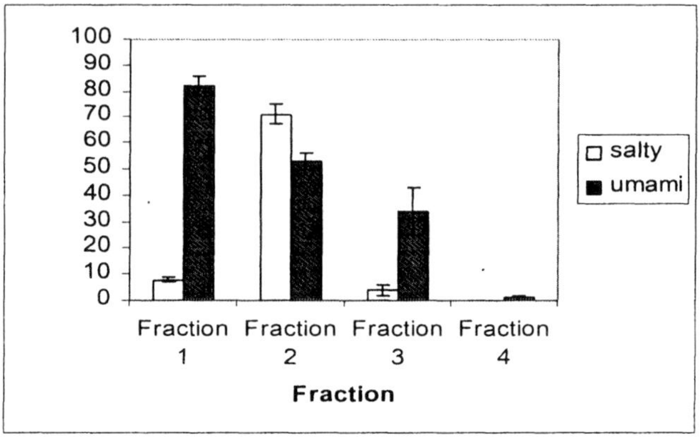

Figure 4. Umami and salty taste intensity of the four fractions of Fraction MW<500 Da. The score range was based on intensity of umami taste or salty taste standard as presented in Table 1. The figures were obtained from the average of 14 trained panelists. Each fraction contained 10.0 mg total soluble peptide/10 mL test solution, except Fraction 2 which contained 1.0 mg/10 mL.

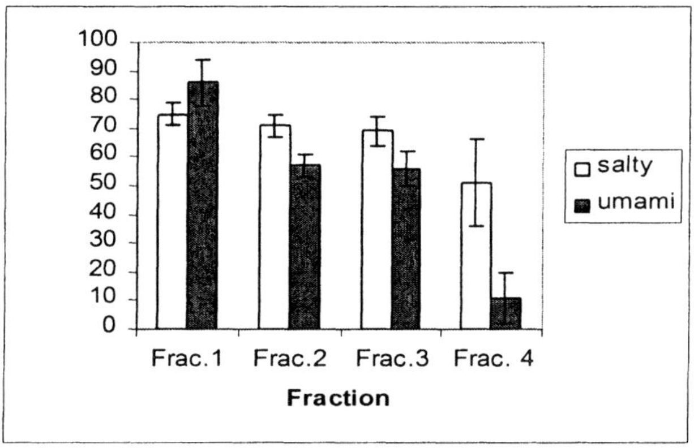

Figure 5. Umami and salty taste intensity of the four fractions of Fraction MW<500 Da. The score range was based on intensity of umami taste or salty taste standard as presented in Table 1. The figures were obtained from the average of 14 trained panelists. Each fraction contained 10.0 mg total soluble peptide/10 mL and salt at concentration 59.9 mg/10 mL tested solution, except Fraction 2 which contained 1.0 mg total soluble peptide/10 mL.

relatively high total glutamic acid content, but a relatively low free glutamic acid content. It may also have more peptides containing non-aromatic amino acids. It can be inferred that several peptides present in Fraction 2 may play a role in high intensity umami taste.

References

1. Adler-Niesen, J. *J. Agric. Food Chem. 1979, 27, 1256-1262.*
2. Apriyantono, A.; Wiratma, E.; Husain, H.; Nurhayati; Lie, L.; Judoamidjojo, M.; Puspitasari-Nienaber, N. L.; Budiyanto, S.; Sumaryanto, H. In *Flavour Science: Recent Developments*; Taylor, A.J., and Mottram, D.S., Eds.; The Royal Society of Chemistry: Cambridge, **1996**; pp 62-65.
3. Apriyantono, A.; Husain, H.; Lie, L.; Judoamidjojo, M.; Puspitasari-Nienaber, N. L. In *Flavor Chemistry of Ethnic Foods*; Shahidi, F. and Ho, C. T., Eds.; Kluwer Academic: New York, **1999**; pp 15-32.
4. Apriyantono, A.; Setyaningsih, D.; Hariyadi, P.; Nuraida, L. In *Quality of Fresh and Processed Foods*; Spanier, A.M., Shahidi, F., Ho, C. T., and Braggins, T., Eds.; Kluwer academic: New York, **2002**.
5. Austen, B. In *New Protein Techniques*; Walker, J. M., Ed.; Humana Press.: Clifton, New Jersey, **1988**; Vol. 3.
6. Bartolome, B.; Arribas, V. M.; Pueyo, E.; Polo, M. C. *J. Agric. Food Chem.* **1997**,*4, 3374 - 3381.*
7. Basha, S.M. *J. Agric. Food Chem.* **1997**, *45, 400-402.*
8. Belitz, H.D.; Grosch, W. *Food Chemistry*; 2nd ed.; Springer: Berlin, **1996**; pp 8-39.
9. Boehringer Mannheim. *Methods of Enzymatic Bioanalysis and Food Analysis*; Boehringer Mannheim GmbH Biochemicals: Mannheim, 1995; pp 56-59.
10. Chung, S.Y.; Ullah, A. H.J.; Sanders, T.H. *J. Agric. Food Chem.* **1994**, *42, 623 – 628.*
11. Cliffe, A.J.; Marks, J.D.; Mulholland, F. *International Dairy Journal.* **1993**, *3, 379 – 387.* Abstract in *Food Science and Technology Abstract*; Institute of Food Information Services: Reading.
12. Frazier, R.A.; Ames, J.M.; Nursten, H.E. *Capillary Electrophoresis for Food Analysis*; The Royal Society of Chemistry: Cambridge, UK, **2000**;p 127.
13. Fukushima, D. *J. Am. Oil Chem. Soc.* **1981**, *58, 346 - 354.*
14. Hagel, L. In *Protein Purification*; Janson, J. C. and Ryden, L., Eds.; 2nd ed.; A John Wiley & Sons, Inc., Publ.: New York, **1998**; pp 79-137.
15. Harris, E.L.V. In *New Protein Techniques*; Walker, J.M., Ed.; Humana Press: Clifton, New Jersey, **1988**; Vol. 3, pp 33-47.

16. Kaneko, K.; Tsuji, K.; Kim, C.H.; Otoguro, C.; Sumino, T.; Aida, K.; Kaneda, T.; Sahara, K. *Journal of Japanese Society of Food Science and Technology.* **1994**, *41, 148 – 156.* Abstract in *Food Science and Technology Abstract*; Institute of Food Information Services: Reading.
17. Kirimura, J.; Shimizu, A.; Komizuka, A.; Ninomiya , T.; Katsuya, N. *J. Agric. Food Chem.* **1969**, *17, 689 - 695.*
18. Lee, K.P.D.; Warthesen, J.J. *J. Agric. Food Chem.* 1996, *44: 1058 - 1063.*
19. Noguchi, M.; Arai, S.; Yamashita, M.; Kato, H.; Fujimaki, M. *J. Agric. Food Chem.* **1975**, *23, 49 – 53.*
20. Ohyama, S.; Ishibashi, N.; Tamura, M.; Nishizaki, H.; Okai, H. *Agric. Biol. Chem.* **1988**, *52, 871 – 872.*
21. Nishimura, T.; Kato, H. *Food Rev. Inter.* **1988**, *4, 175 – 194.*
22. Oka, S.; Nagata, K. *Agric. Biol. Chem.* **1974**, *38, 1185 - 1194.*
23. Pavia, D. L.; Lampman, G. M.; Kriz, G. S. *Introduction to Spectroscopy*; 2nd ed.; Harcourt Brace College Publishers: London, **1996**; p 511.
24. Sasaki, M. *J. Agric. Food Chem.* **1996**, *44, 3273 - 3275.*
25. Sommerer, N.; Garem, A.; Molle; Septier, C.; Le Quere, J. L.; Salles, C. In *Food Flavors: Formation, Analysis and Packaging Influences*; Contis, E. T., Ho, C.-T., Mussinan, C. J., Parliment, T. H., Shahidi, F., and Spanier, A.M., Eds.; Elsevier: Amsterdam, **1998**; pp 207-217.
26. Spanier, A. M.; Bland, J.M.; Flores, M.; Bystricky, P. In *Chemistry of Novel Foods*; Spanier, A. M., Tamura, M., Okai, H. and Mills, O., Eds.; Allured Publishing Co.: Carol Stream, IL, **1997**; pp 45-66.
27. Takeuchi, T. *Journal of Fermentation Technology.* 1974, *52, 256 – 267.* Abstract in *Food Science and Technology Abstract*; Institut of Food Information Services: Reading.
28. Tamura, M.; Nakatsuka., T.; Tada, M.; Kawasaki, Y.; Kikuchi, E.; Okai, H. *Agric. Biol. Chem.* **1989**, *53 , 319 - 325.*
29. Valentin, J.; Guillard, A.S.; Septier, C.; Salles, C.; Le Quere, J.L. In *Food Flavors: Formation, Analysis and Packaging Influences*; Contis, E.T., Ho, C.-T., Mussinan, C.J., Parliment, T.H., Shahidi, F., and Spanier, A.M. , Eds.; Elsevier: Amsterdam, **1998**; pp 195-205.
30. Walker, J.M. *New Protein Techniques*; Humana Press: Clifton, New Jersey, **1988**.
31. Yamasaki, Y.; Maekawa, K. *Agric. Biol. Chem.* **1978**, *42, 1761-1765.*
32. Yamasaki, Y.; Maekawa, K. *Agric. Biol. Chem.* **1980**, *44, 93 – 97.*

Chapter 13

Synthesis, Structure, and Activity of Novel Glycoconjugates Exhibiting Umami Taste

F. Robert[1], I. Blank[1], L. B. Fay[1] E. Beksan[2], T. Hofmann[3], and P. Schieberle[2]

[1]Nestec Ltd., Nestlé Research Center, Vers-chez-les-Blanc, 1000 Lausanne 26, Switzerland
[2]Deutsche Forschungsanstalt für Lebensmittelchemie, Lichtenbergstrasse 4, D–85748 Garching, Germany
[3]Institut für Lebensmittelchemie, Corrensstrasse 45, Westfälische Wilhelms-Universität Münster, D–48149 Münster, Germany

Two new umami-tasting compounds were prepared by a Maillard-mimetic synthesis approach. The chemical structures of these novel glycoconjugates were elucidated as *N*-(D-glucos-1-yl)-L-glutamate and the corresponding Amadori compound *N*-(1-deoxy-D-fructos-1-yl)-L-glutamate. Their stability was studied by ^{13}C NMR spectroscopy, showing that the Amadori compound was much more stable than the *N*-glucosyl derivative, which limits the use of *N*-glucosyl glutamate in food applications.

Introduction

Since a few years, the so-called umami taste is accepted as the fifth basic taste quality along with the taste modalities sweet, sour, salty, and bitter. This is mainly due to the identification of G protein-coupled receptors for glutamate such as mGluR4 (*1*), which is a truncated form of the metabotropic glutamate receptor 4 in the brain, and the heteromer T1R1+3 receptor (*2*), which was reported to be a a broadly tuned receptor stimulated by many L-amino acids, in particular also by glutamate.

Monosodium glutamate (MSG) is the best known compound eliciting umami taste. This unique taste sensation was first described by Ikeda (*3*). Other compounds with similar sensory characteristics belong to the group of purine-5'-nucleotides, such as inosine-5'-monophosphate (IMP) and guanosine-5'-monophosphate. These compounds occur in many savory foods such as meat, fish, seafood, and mushrooms (*4*, *5*). An interesting property of umami compounds is their mutual taste synergism. The synergistic effects between MSG and IMP have been investigated by Yamaguchi and coworkers (*6*, *7*).

In this paper, we focus on synthesis and structure elucidation of novel glycoconjugates, e.g. *N*-glucosyl glutamate and the Amadori compound deoxyfructosyl glutamate, which represent a new class of umami-tasting compounds. Their chemical stability was studied by ^{13}C NMR measurements.

Experimental

Synthesis of dipotassium N-(D-glucos-1-yl)-L-glutamate (DPGG). A solution of potassium hydroxide (167 mmol) in anhydrous methanol (100 mL, H_2O < 0.02 %, Merck, Darmstadt, Germany) was added to anhydrous monopotassium glutamate (167 mmol, Sigma, Buchs, Switzerland) and stirred until the amino acid was completely dissolved. Methanol (300 mL) and anhydrous D-glucose (100 mmol, Aldrich, Buchs, Switzerland) were added and the suspension was stirred and heated under reflux for 1 h at 75°C. The reaction mixture was cooled to ambient temperature and concentrated to 100 mL under vacuum. The surplus of amino acid was removed by filtration and the filtrate containing the product was cooled to -20 °C. The *N*-glucoside was obtained by adding acetone (Merck) dropwise to the filtrate. The sample was purified by a 2-fold recrystallization from methanol/acetone according to the procedure described above. Finally, the product was suspended in acetone, homogenized by ultrasound and dried under a nitrogen flow to furnish 12.7 g of a white, hygroscopic powder (33 % yield, >90% purity with remaining methanol).

Synthesis of N-(1-deoxy-D-fructos-1-yl)-L-glutamic acid (Fru-Glu). DPGG (3.04 g, 7.9 mmol) was refluxed for 30 min in anhydrous methanol (30 mL) containing acetic acid (1 mL). The solution was cooled to 0 °C prior to the addition of acetone (20 mL). The precipitate was filtered under reduced pressure and dried under high vacuum to yield a pale yellow powder. It was further purified by dissolving it in water and lyophilizing overnight to obtain 1.23 g of a pale brown powder (52 % yield, >90% purity with remaining glutamate).

Mass Spectrometry. Direct infusion mass spectrometry was carried out on a Micromass (Manchester, UK) Quattro-LC triple quadrupole mass spectrometer equipped with a "Z-Spray" electrospray ion source. The electrospray capillary voltage was set to 4.0 kV, the source block temperature to 80 °C. The cone gas was operated at 70 L/h, desolvation gas at 550 L/h and the desolvation temperature to 150 °C. Data were acquired in negative mode from 50 to 800 Da with a dwell time of 1 s. Data acquisition was performed using the software package MassLynx 3.4 of Micromass. The samples were introduced at 10 µL/min in a mixture of methanol/water (1/1, v/v).

Nuclear Magnetic Resonance (NMR) Spectroscopy. The samples for NMR spectroscopy were prepared in WILMAD 528-PP 5 mm Pyrex NMR tubes, using heavy water as solvent (0.7 mL, 99.8%, Fluka, Buchs Switzerland). The NMR spectra were acquired on a Bruker AM-360 spectrometer, equipped with a quadrinuclear 5 mm probe head, at 360.13 MHz for ^{1}H and at 75.56 MHz for ^{13}C. All shifts are cited in ppm. One-dimensional ^{1}H-NMR, ^{13}C-NMR, Distortionless Enhancement by Polarisation Transfer (DEPT-135), and two-dimensional COSY and HETCOR spectra were acquired as described earlier using standard conditions (*8*).

Stability of dipotassium N-(D-Glucos-1-yl)-L-glutamate (DPGG). This was performed at different pH values in aqueous solutions by ^{13}C NMR measurements. Solutions were prepared with about 50 mg of DPGG in D_2O/H_2O (1 + 5, v/v, 0.7 mL). If necessary, the pH was adjusted with sulphuric acid. The sample was transferred into a NMR tube and the NMR experiment recorded over a defined period of time.

Results and Discussion

Synthesis

N-(D-Glucos-1-yl)-L-glutamate was obtained by a new and simple synthetic pathway (Figure 1) in two steps starting from monopotassium L-glutamate, which was first solubilized in anhydrous methanol in the presence of KOH and then coupled with D-glucose under reflux conditions resulting in the dipotassium salt of *N*-(D-glucos-1-yl)-L-glutamic acid (DPGG) in about 40 % overall yield after purification. It is important to work under anhydrous conditions to avoid hydrolysis of the *N*-glucoside or its rearrangement to the Amadori compound, i.e. *N*-(1-deoxyfructos-1-yl)-L-glutamate (Fru-Glu). On the other hand, the latter reaction is a convenient approach to obtain Fru-Glu, another glutamate derivative. Acidic conditions allowed dehydration and enolization leading to the Amadori rearrangement product Fru-Glu in about 50 % yield. Both compounds were found to elicit an umami-like taste.

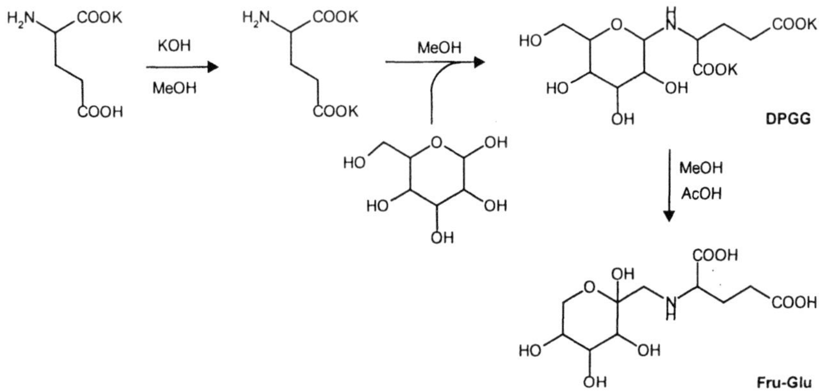

Figure 1. Synthesis pathway to dipotassium N-(D-glucos-1-yl)-L-glutamate (DPGG) and N-(1-deoxyfructos-1-yl)-L-glutamic acid (Fru-Glu)

From the synthesis point of view, the reaction of reducing sugars and amino acids in anhydrous methanol under basic pH conditions offers an elegant synthetic approach towards *N*-glycosides. This has first been described by *Weitzel* and coworkers (*9*). The yields obtained in this work for DPGG were typically 30-40 %. Anhydrous conditions reduce decomposition by hydrolysis and the rearrangement reaction to the corresponding Amadori compound Fru-Glu. The starting materials are cheap and the reaction is rather simple.

Structure elucidation

N-(D-Glucos-1-yl)-L-glutamate (dipotassium salt, DPGG). The compound was characterized by various NMR and MS techniques (Table I). Not only ^{1}H- and ^{13}C-NMR experiments were applied, but also DEPT (θ = 135°) and 2-D NMR sequences (COSY and HETCOR) for unequivocal assignments of C and H atoms. The ^{1}H-NMR spectrum (data not shown) did not allow full interpretation of the signals due to signal overlapping. However, this experiment was useful to estimate the residual methanol amount present in the sample, which was found to be significant and accounted for up to about 10 %.

^{13}C NMR in combination with DEPT-135 resulted in 11 signals, of which 6 were CH or CH_3 groups, 3 CH_2 groups, and 2 quaternary carbons (Figure 2). The chemical shifts of the two carboxylate groups (185.6 and 185.0 ppm) showed clearly the anionic form.

Table I. NMR data (*δ* / ppm) of dipotassium *N*-(D-glucos-1-yl)-L-glutamate.

^{1}H-NMR (360 MHz, D_2O)	^{13}C-NMR (90 MHz, D_2O)
1.82 (m, 2H, $C_{3'}$); 2.22 (m, 2H, $C_{4'}$); 3.20 (dd, ^3J = 8.8 Hz, 1H, C_2); 3.33-3.37 (m, 2x 1H, C_4 and C_5); 3.41 (m, 1H, $C_{2'}$); 3.45 (m, 1H, C_3); 3.67 (dd, ^2J = 12.3 Hz, ^3J = 5.4 Hz, 1H, C_6); 3.85 (dd, ^2J = 12.3 Hz, ^3J = 0.8 Hz, 1H, C_6); 3.89 (d, ^3J = 8.8 Hz, 1H, C_1).	33.1 ($C_{3'}$); 37.1($C_{4'}$); 63.0 ($C_{2'}$); 63.9 (C_6); 72.7 (C_5); 75.8 (C_2); 79.5 (C_3); 79.7 (C_4); 91.7 (C_1); 185.0 (CO_2); 185.6 (CO_2).

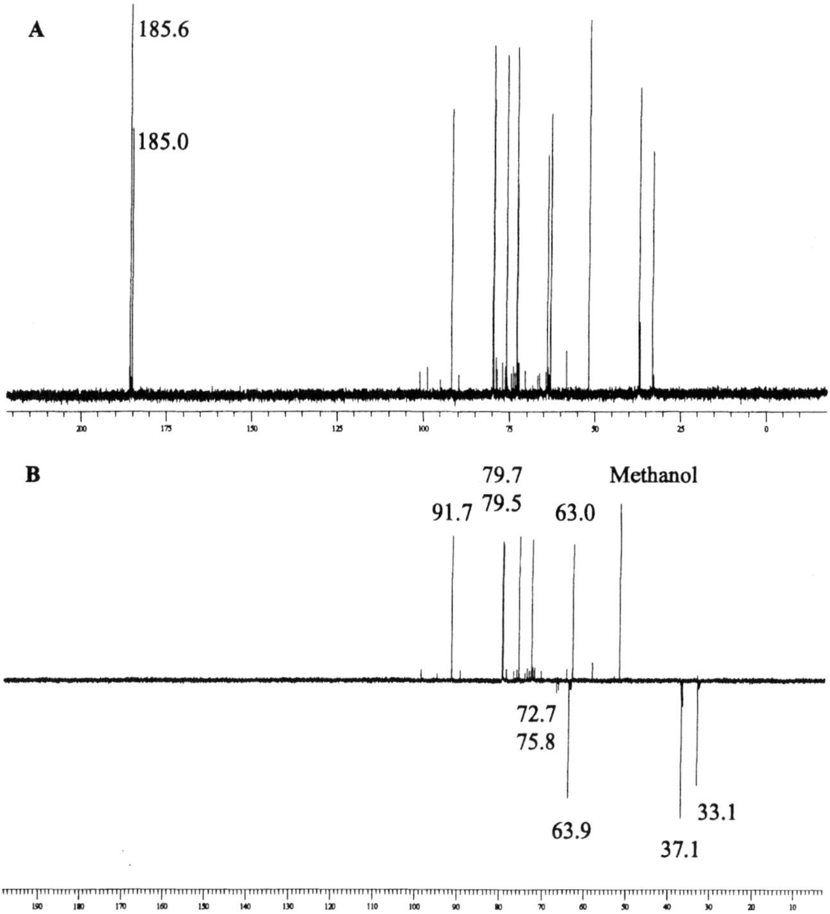

Figure 2. ^{13}C-NMR (A) and DEPT-135 (B) spectra of dipotassium N-(D-glucos-1-yl)-L-glutamate (chemical shifts in ppm).

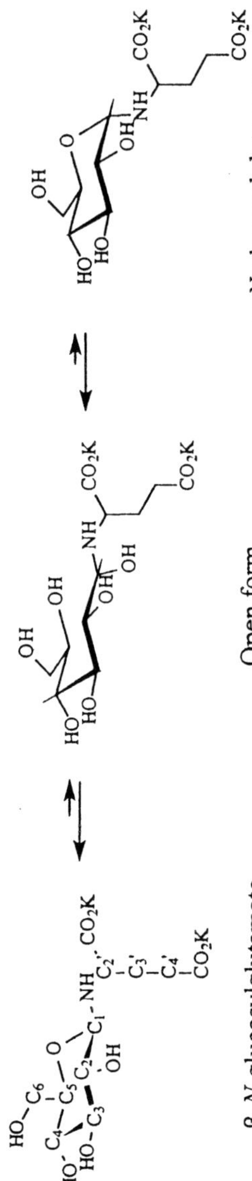

β-*N*-glucosylglutamate ⇌ Open form ⇌ α-*N*-glucosylglutamate

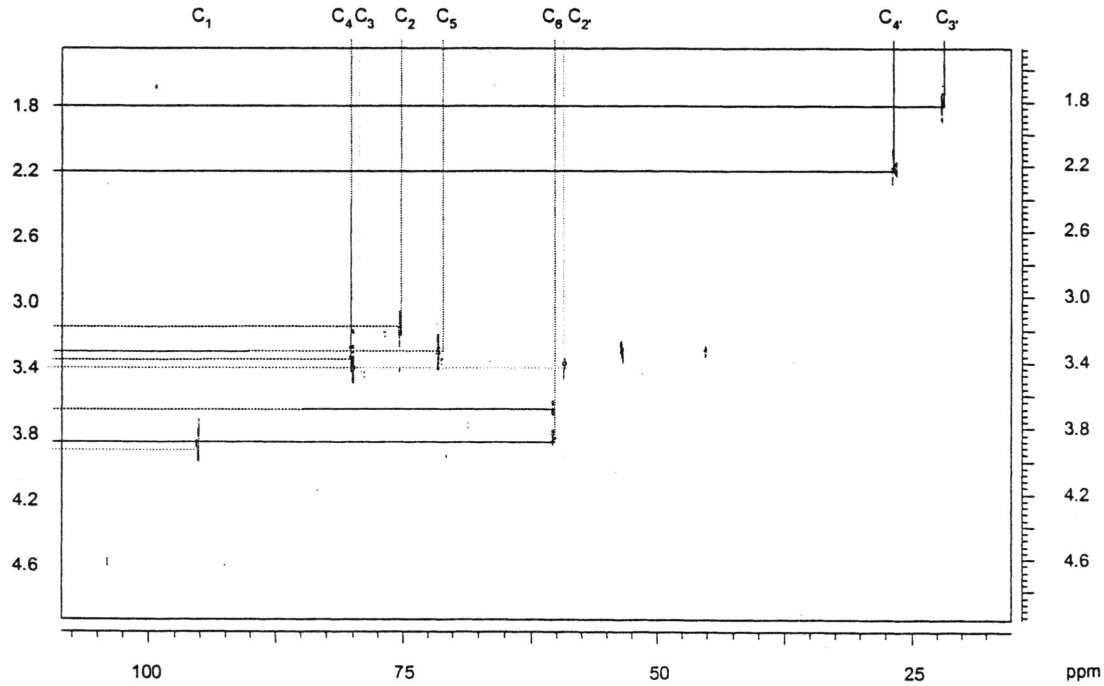

Figure 3. HETCOR NMR of dipotassium N-(D-glucos-1-yl)-L-glutamate.

The signal at 91.7 ppm is the anomeric carbon, which is close to an oxygen and a nitrogen atom. This signal proves that the structure is of an *N*-glucosyl type. The five CHOH of the sugar moiety are shifted between 79.7 and 72.7 ppm whereas the CH_2OH group is at 63.9 ppm. The signals at 63.0, 37.1 and 33.1 ppm correspond to the CH and the two CH_2 of the glutamate backbone, respectively. The presence of only 3 methylene protons ruled out the Amadori structure of the compound synthesized. The signal at 91.7 ppm for the anomeric carbon suggested the structure of an *N*-glycoside, which is compatible with the mechanism of the Maillard reaction and the conditions used in the synthesis.

Thanks to HETCOR (Figure 3) and COSY (Figure 4) experiments, the complete set of ^{1}H and ^{13}C signals could be assigned as shown in Table I. HETCOR allowed to define the anomeric proton at 3.89 ppm, which coupled with the carbon at 91.7 ppm. From this signal, all the other proton shifts of the sugar skeleton could be attributed with the COSY experiment. The low-field methylenes of the amino acid part were also distinguished. Moreover, the anomeric proton shifted at 3.89 ppm indicates a doublet with a ^{3}J constant coupling of 8.8 Hz characteristic of a *trans* axial-axial coupling, thus suggesting the β-glucopyranose as the main form.

Negative electrospray mass spectrometry ESI (-) MS measurements were performed to confirm the molecular weight and the structure of the compound. In diluted solutions (10 ng/μL), the ion at *m/z* 326, corresponding to the molecular ion (Figure 5) of the open-chain form shown below, was found to be associated with an ion at *m/z* 308. At higher concentration (200 ng/μL), however, only *m/z* 308 showed up (data not shown), which suggests the cyclic form of the molecule. MS/MS experiments of the peaks at *m/z* 308 and 326 confirmed the presence of both compounds occurring individually in the synthesized sample (Table II). The molecular ion at *m/z* 326 resulted only in *m/z* 146 as daughter ion, thus suggesting that *m/z* 308 was not formed by dehydration of *m/z* 326. In agreement with that, parents measurement of *m/z* 308 did not yield in *m/z* 326, which confirms the presence of the cyclic and the open-chain form in equilibrium as shown below.

N-(1-Deoxy-D-fructos-1-yl)-L-glutamic acid (Fru-Glu). The Amadori-type structure of Fru-Glu was mainly evidenced by ^{13}C-NMR/DEPT data (Table III) indicating 4 methylene carbons out of 11 signals, instead of 3 for DPGG. In agreement with that, C1 in Fru-Glu was clearly defined as a CH_2 group, instead of a CH group in DPGG. Furthermore, C2 in Fru-Glu was represented by a quaternary carbon, and in DPGG by a methine group.

The reaction of amino compounds (e.g. amino acids) with reducing sugars leads first to *N*-glycosides, which then can form "Amadori" rearrangement products (from aldoses) as first relatively stable intermediates in the course of the so-called "Maillard reaction", the non-enzymatic browning reaction (review in *10*, *11*). Basically, it is possible to stop the reaction at the level of *N*-glucosyl glutamate. Upon acidification and heat treatment, however, the reaction can also

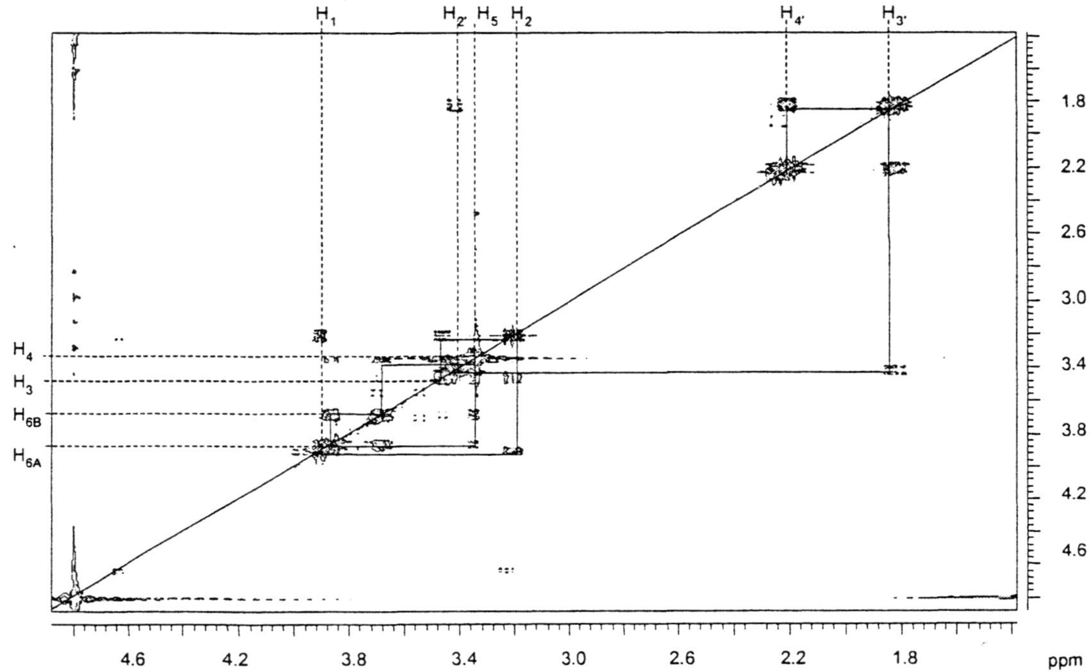

Figure 4. COSY NMR of dipotassium N-(D-glucos-1-yl)-L-glutamate.

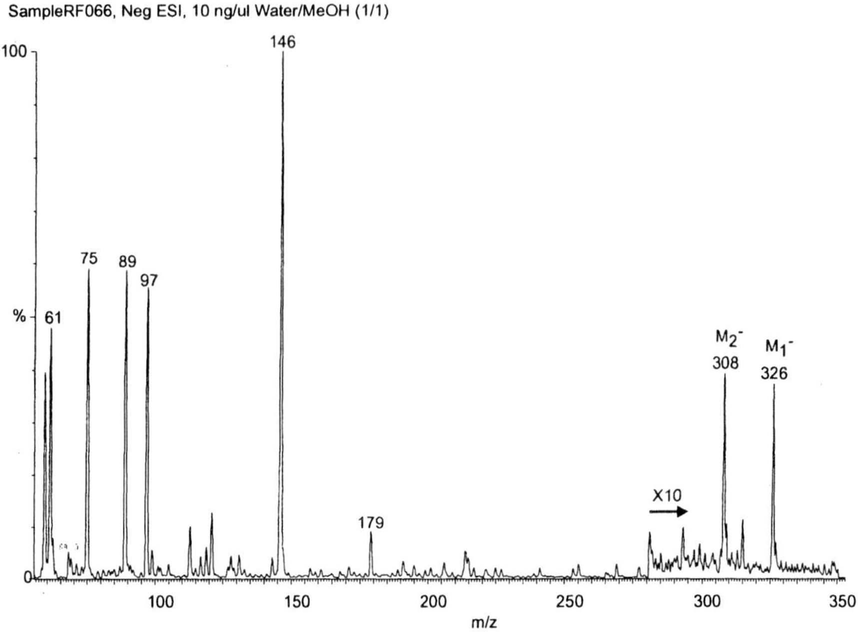

Figure 5. MS-ESI (-) of dipotassium N-(D-glucos-1-yl)-L-glutamate.

Table II. MS spectral data of dipotassium *N*-(D-glucos-1-yl)-L-glutamate.

Compound	Parent ion (*m/z*)	Daughter fragments (Intensity %)
Dipotassium *N*-(D-Glucos-1-yl)-L-glutamate, open-chain form	326 $[M-H]^-$	146 (100)
Dipotassium *N*-(D-Glucos-1-yl)-L-glutamate, cyclic form	308 $[M-H]^-$	188 (100), 264 (5), 170 (5), 144 (5)

Table III. NMR data (δ / ppm) of *N*-(1-deoxyfructos-1-yl)-L-glutamic acid.

^{1}H-NMR (360 MHz, D_2O)	^{13}C-NMR (90 MHz, D_2O)
2.04-2.10 (m, 1H, CH_2, $C_{3'}$); 2.39-2.55 (m, 3H, CH_2, $C_{3'}$); 3.13 (d, ^{2}J = 14.5 Hz, 1H, CH_2, C_1); 3.42-3.98 (m, 6H).	28.51(CH_2); 37.0 (CH_2); 55.5 (CH_2, C_1); 66.2 (CH); 66.7 (CH_2, C_6); 71.8, 72.2, 72.7 (3 CH, C_3, C_4, C_5); 98.3 (C_q, C_2); 176.3 (C_q, CO_2); 183.9 (C_q, CO_2).

proceed to the Amadori compound *N*-(1-deoxy-D-fructos-1-yl)-L-glutamate (Fru-Glu) *via* the Amadori rearrangement (*12*). This is a new synthetic pathway to obtain Amadori compounds in fairly good yields.

DPGG seems to be a new compound, not yet reported in the literature. However, the chemical class of *N*-glycosides of glucose and glutamate is known. The synthesis of glucosylglutamate and derivatives has been described earlier (*9, 13*), but not its occurrence in food. On the contrary, the Amadori compound Fru-Glu is a known consituent of food. High amounts (up to 3.6 g/100 g dry matter) of Fru-Glu have been reported in dried tomates (*14*). However, it was also found in other dried vegetables such as celery, asparagus, cauliflower, carrots, and red pepper as well as dark malts (*14, 15*).

Stability

First indication for the instability of DPGG in aqueous solutions was found by analytical measurements. Analysis by HPLC indicated glutamate as the major product. On the other hand, analysis of DPGG by the high performance anion exchange chromatography revealed glucose as the major product (data not shown). These findings indicated that DPGG readily hydrolyses to glucose and glutamate. A series of ^{13}C NMR experiments were performed to get an insight into the chemical stability of DPGG in aqueous solutions by varying the parameters temperature, time and pH. The hydrolysis of DPGG was monitored in the range of 90 to 100 ppm. The release of free glucose was indicated by the signals at 95.0 and 98.8 ppm, which correspond to the α and β anomeric C-atoms, respectively.

The influence of pH on the stability of DPGG was studied in aqueous solutions under acidic, neutral, and alkaline conditions by performing ^{13}C NMR measurements 15 min after sample preparation (Figure 6).

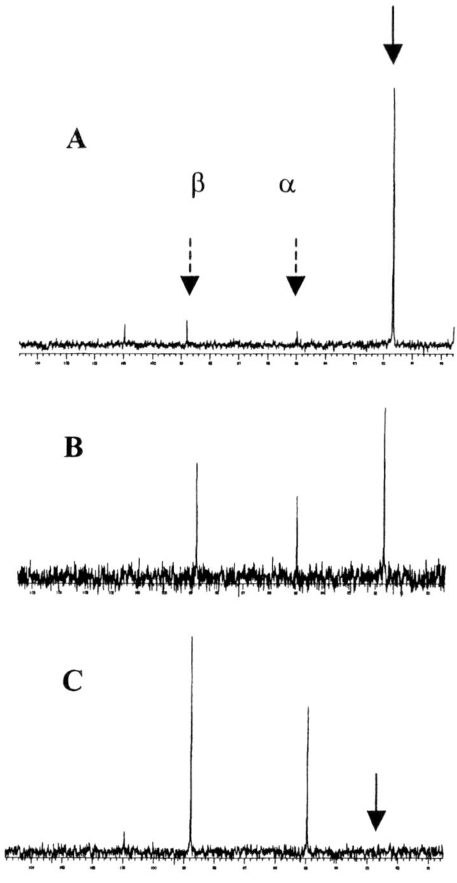

*Figure 6. Stability of dipotassium N-(D-glucos-1-yl)-L-glutamate in D_2O at pH 9.3 (**A**), pH 6.5 (**B**), and pH 2.3 (**C**) after 15 min of sample preparation. The full flash indicates the signal at 91.7 ppm for DPGG, the dotted flashes represent the anomeric C-atoms of free glucose (95.0, 98.8 ppm).*

As shown in Figure 6, DPGG is stable at pH 9.3 as the anomeric C-atoms of free glucose (dotted flash) are not pronounced, indicating that no or only little hydrolysis occurred. At pH 6.5, however, DPGG hydrolyses to a certain extent as indicated by the two anomeric C-atoms of glucose liberated upon hydrolysis. Rapid hydrolysis of DPGG was observed at pH 2.3, and consequently no signal for DPGG could be found. Under acidic conditions, this takes place instantaneously. The instability of DPGG under acidic to neutral pH conditions may be the reason for the few literature data available on such compounds.

In general, N-glycosides of amino acids are known to be very unstable in aqueous solutions (9). The N-glycosidic bond is particularly labile under acidic and neutral pH conditions (16, 17). Rapid hydrolysis shifts the equilibrium to the adducts, i.e. reducing sugars and amino acids. With increasing pH, however, N-glycosides become more stable and can be isolated either by crystallization with methanol/acetone or as bimolecular complex species thanks to polyvalent alcaline or transition metals (13).

The high instability in aqueous solutons is also valid for N-glucosyl-L-glutamate. Therefore, it is not surprising that it has not yet been reported in food products. However, the presence of N-glucosyl derivatives of amino acids as intermediates in the Maillard reaction is well established and generally accepted (10). Depending on the recation conditions, N-glucosyl-L-glutamate may decompose via hydrolysis giving rise to glucose and glutamate or rearrange to the corresponding Amadori compound. On the contrary, reaction of glucose with amino acids in anhydrous methanol, as shown with glutamate in this study, results in N-(D-glucos-1-yl)-L-glutamate with fairly good yields.

In contrast to *N*-glucosyl-L-glutamate, the Amadori compound Fru-Glu was found to be rather stable (Figure 7). No increase of the anomeric C-atoms of free glucose (95.0, 98.8 ppm) could be observed at pH 4, even after 3 days of storage at room temperature. The glucose indicated by the dotted flasches correspond to trace amounts remaining in the Fru-Glu sample after purification.

In conclusion, two new umami-tasting compounds were prepared and their chemical structures elucidated, namely the *N*-glycoside dipotassium *N*-(D-glucos-1-yl)-L-glutamate and the corresponding Amadori compound *N*-(1-deoxy-D-fructos-1-yl)-L-glutamic acid. To the best of our knowledge, these types of glycoconjugates in general and *N*-glucosyl glutamate and *N*-deoxyfructosyl glutamate in particular have never been reported as taste active compounds having umami-like properties. Therefore, glycosylglutamates represent a new class of potentially umami-type tasting compounds showing similar properties as the umami reference compound MSG. ^{13}C NMR was found to be a rapid and reliable method to study the stability of glyco-conjugates, which are known to be rather unstable. This can be performed in D_2O or D_2O/H_2O mixtures if buffers are to be used. The particular importance of NMR is due to the lack in alternative analytical tools to monitor the labile *N*-glycosides.

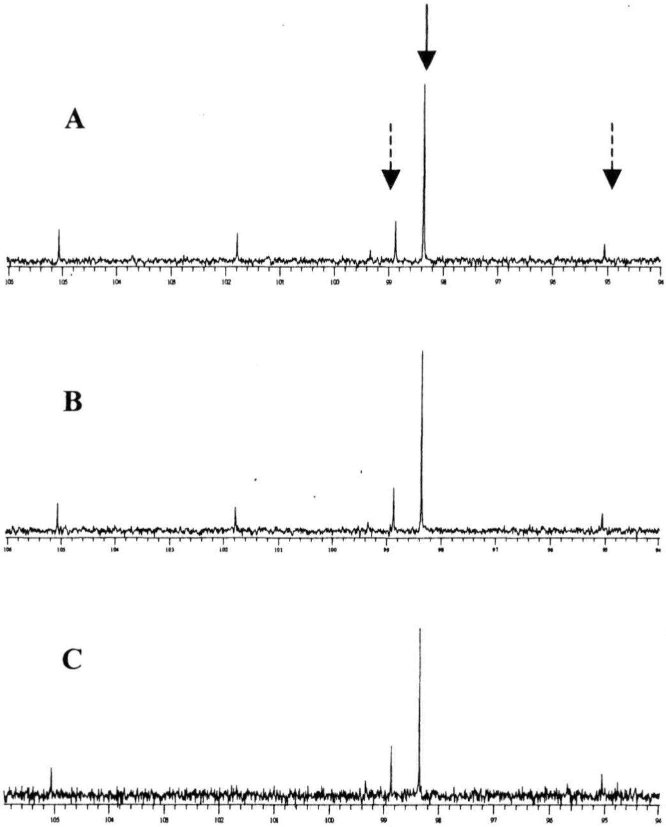

*Figure 7. Stability of N-(1-deoxy-D-fructos-1-yl)-L-glutamic acid in water at pH 4.0 after 3 h (**A**), 24 h (**B**), and 96 h (**C**). The full flash indicates the C2 signal at 98.3 ppm, the dotted flashes represent the anomeric C-atoms of free glucose (95.0, 98.8 ppm).*

References

1. Chaudari, N.; Landin, A. M.; Roper, S. D. *Nat. Neurosci.* **2000**, *3*, 113-119.
2. Nelson, G.; Chandrashekar, J.; Hoon, M. A.; Feng, L.; Zhao, G.; Ryba, N. J. P.; Zuker, C. S. *Nature* (London) **2002**, *416*(6877), 199-202.
3. Ikeda, K. *J. Tokyo Chem. Soc.* **1909**, *30*, 820-826.
4. Kuninaka, A. In: *Symposium on Foods: The Chemistry and Physiology of Flavors*; Schultz, H. W.; Day, E. A.; Libbey, L. M., Eds.; AVI Publishing Company: Westport, CT, 1967; pp 515-535.
5. Yamaguchi, S. *J. Food Sci.* **1967**, *32*, 473-478.
6. Yamaguchi, S.; Yoshikawa, T.; Ikeda, S.; Ninomiya, T. *J. Food Sci.* **1971**, 36, 1761-1765.
7. Ninomiya, K. *Food Rev. Int.* **2002**, *18*, 23-38.
8. Lin, J.; Welti, D. H.; Arce Vera, F.; Fay, L. B.; Blank, I. *J. Agric. Food Chem.* **1999**, *47*, 2813 2821.
9. Weitzel, G.; Geyer, H.-U.; Fretzdorff, A.-M. *Chem Ber.* **1957**, *90*, 1153-1161 (in German, *Chem. Abstr.* 1958, *52*, 11741).
10. Ledl, F.; Schleicher, E. *Angew. Chem. Int. Ed. Engl.* **1990**, *29*, 565-594.
11. Yaylayan, V. A.; Huyghues-Despointes, A. *Crit. Rev. Food Sci. Nutr.* **1994**, *34*, 321-369.
12. Hodge, J.E. *Advan. Carbohyd. Res.* **1955**, *10*, 169-205.
13. Chen, J.; Pill, T.; Beck, W. *Z. Naturforsch.* **1989**, *44b*, 459-464 (in German, *Chem. Abstr.* 1990, *112*, 139749a).
14. Eichner, K.; Reutter, M; Wittmann, R. In *The Maillard Reaction in Food Processing, Human Nutrition and Physiology*; Finot, P.A., Aeschbacher, H.U., Hurrel, R.F., Liardon, R., Eds.; Birkhauser Verlag: Basel, 1990; pp 63-77.
15. Heinzler, M.; Eichner, K. *Z. Lebensm. Unters. Forsch.* **1991**, *192*, 24-29 (in German; *Chem. Abstr.* 1991, *115*, 27981a).
16. Von Euler, H.; Brunius, E. *Chem. Ber.* **1926**, *59*, 1581-1585 (in German, *Chem. Abstr.* 1926, *20*, 3826).
17. Paulsen, H.; Pflughaupt, H. In *The Carbohydrates - Chemistry and Biochemistry*; Pigman, W.; Horton, D., Eds.; Academic Press: New York, NY, 1980; Vol 1B, pp 881-927.

Chapter 14

Taste-Active Glycoconjugates of Glutamate: New Umami Compounds

Hedwig Schlichtherle-Cerny, Michael Affolter, and Christoph Cerny

[1]**Nestlè Research Center, Nestec Ltd., 1000 Lausanne 26, Switzerland**

Hydrolyzed plant proteins are widely used as ingredients in culinary products for their glutamate-like "umami" taste. The comparison of the taste profiles of three different wheat gluten hydrolysates revealed that the enzymatic hydrolysate of acid-deamidated wheat gluten elicited an intense glutamate-like taste. The hydrolysate was analyzed by "LC-tasting", a technique comprising stepwise fractionation of a food by different chromatographic methods, such as gel permeation chromatography and RP-HPLC, followed by the sensory evaluation of the obtained fractions after lyophilization. The chemical analysis of the most intense glutamate-like subfraction by hydrophilic interaction liquid chromatography coupled to electrospray ionization mass spectrometry (HILIC-ESI-MS) revealed many hydrophilic glutamyl di- and tripeptides, as well as the presence of different glycoconjugates of glutamate, glutamine, and lysine. The most abundant Amadori compound, *N*-(1-deoxy-fructos-1-yl) glutamate, was identified as eliciting an intense umami taste. The glycoconjugates of glutamate were identified as a new class of compounds imparting umami taste to foods.

The Japanese word „umami“ means delicious and is used as a synonym for the characteristic sensory properties of monosodium glutamate (MSG) and certain purine-5'-nucleotides like adenosine-5'-monophosphate (5'-AMP), inosine-5'-monophosphate (5'-IMP) and guanosine-5'-monophosphate (5'-GMP) (*1, 2*). Apart from L-glutamic acid its homologue L-aspartic acid as well as lactic and succinic acid elicit umami taste (*3-5*). Also *N*-lactoyl glutamate, the condensation product of lactic and glutamic acid, has glutamate-like organoleptic properties (*6*). Figure 1 gives an overview of the chemical structures of these umami compounds.

Hydrophilic low-molecular weight peptides, in particular those composed of glutamic acid and other hydrophilic amino acids, have also been described as eliciting bouillon-like taste qualities (*7-9*). Van den Oord and van Wassenaar (*10*), however, generally question the existence of umami peptides.

Systematic approaches are necessary to evaluate the taste-active compounds of foods. "LC-tasting" is a combination of instrumental and sensory analysis. An extract is stepwise fractionated by different chromatographic methods (e.g. gel permeation chromatography GPC, RP-HPLC), the interesting fractions are detected by their sensory evaluation after removing solvents and characterized by analytical techniques such as liquid chromatography mass spectrometry (LC-MS). In analogy to the "aroma extraction dilution analysis" (*11*), "taste dilution analysis" was recently proposed to screen taste-active molecules (*12*).

The present work shows the application of LC-tasting to a wheat gluten hydrolysate and the identification of new taste-active compounds in one of the subfractions.

Experimental

Materials

All chemicals and solvents used in the study were commercially available and of analytical or HPLC grade. *N*-(1-deoxy-fructos-1-yl) glutamate, synthesized according to (*13*) was provided by F. Robert (Nestlé Research Center, Lausanne).

Wheat gluten hydrolysates were prepared according to Schlichtherle-Cerny and Amadò (*14*). For wheat gluten hydrolysate 1 (WGH-1) a proteinase and peptidase cocktail (Flavourzyme™, Novozymes, Bagsværd, Denmark) was used for hydrolysis. WGH-2 was produced by employing Flavourzyme™ together with a glutaminase (Glutaminase C-100, Valley Research, South Bend, IN). WGH-3 was obtained by enzymatic hydrolysis with Flavourzyme™ of partially acid-deamidated wheat gluten.

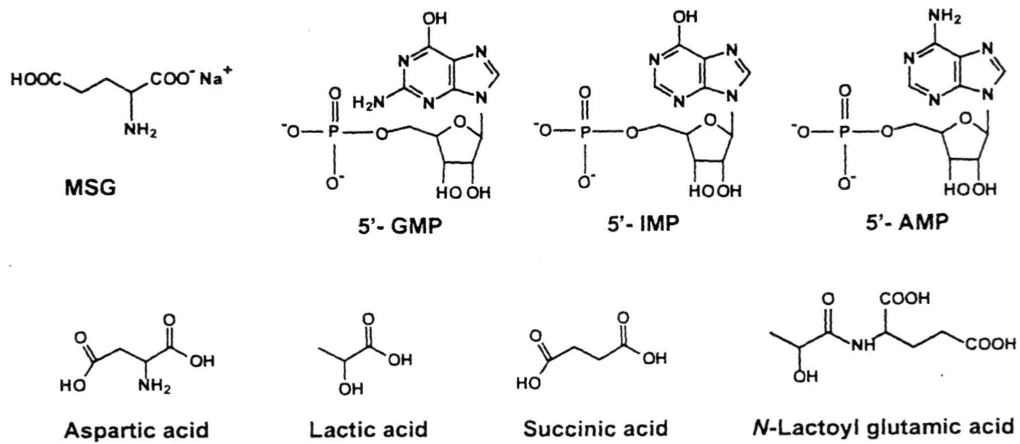

Figure 1. Compounds with umami taste

Instrumental Analysis

An ultrafiltrate of WGH-3 (molecular weight MW < 3 kDa) was fractionated by gel permeation chromatography (GPC) on Sephadex G-10 using 15 % ethanol in deionized water as eluent.

RP-HPLC subfractionation was performed with ammonium acetate buffer (pH 6.0) and methanol on a reversed phase C-18 stationary phase (*14*).

Hydrophilic interaction liquid chromatography - electrospray ionization mass spectrometry (HILIC-ESI-MS) of RP-HPLC subfractions was done on a TSKGel Amide-80 column (Tosoh Biosep, Tokyo, Japan) using ammonium acetate buffer (6.5 mM, pH 5.5, in 90 % acetonitrile) as solvent A, and ammonium acetate buffer (7.2 mM, pH 5.5, in 60 % acetonitrile) as solvent B. The gradient was 13-40 % water in 90 minutes. ESI-MS analysis was performed using a LCQ ion trap mass spectrometer (Thermo Finnigan, San José, CA) in the positive ionization mode (ESI source 4.0 kV, capillary heater 200 °C). MS-MS and MS^3 spectra were acquired with a relative collision energy of 35 %.

Sensory Analysis

Spray-dried wheat gluten hydrolysates (10 g/L) and freeze-dried GPC and RP-HPLC fractions, corresponding to 10 g/L WGH-3, were re-dissolved in deionized water. The intensities of the taste attributes sweet, sour, salty, umami and bitter, judged by 8 trained panelists on a 100-mm line scale ranging from 0 (= absent) to 100 (= very strong), were averaged.

N-(1-deoxy-D-fructos-1-yl)-L-glutamate (Fru-Glu, 10 mmol/L) and MSG (10 mmol/L) were tested sensorially in water and in a bouillon base by 6 trained panelists. The solutions were adjusted to pH 6.0 by addition of sodium hydroxide solution or hydrochloric acid. The aqueous samples were presented in coded beakers at room temperature (22 ± 2 °C), while the bouillon samples were tested at 65 °C. The panelists were asked to describe the taste of the samples using their own descriptors and to compare the taste intensities of the solutions.

Results and Discussion

Three different wheat gluten hydrolysates were investigated. The production scheme is shown in Figure 2.

The first hydrolysate, WGH-1, was hydrolyzed by Flavourzyme™ alone, a commercial enzyme cocktail of endo- and exopeptidases. Since wheat gluten contains about 35 % glutamine, the hydrolysate is rich in glutaminyl peptides and free glutamine. The second hydrolysate, WGH-2, was produced with Flavourzyme™ and additionally a glutaminase to convert released glutamine into free glutamic acid. For WGH-3 wheat gluten was first treated with

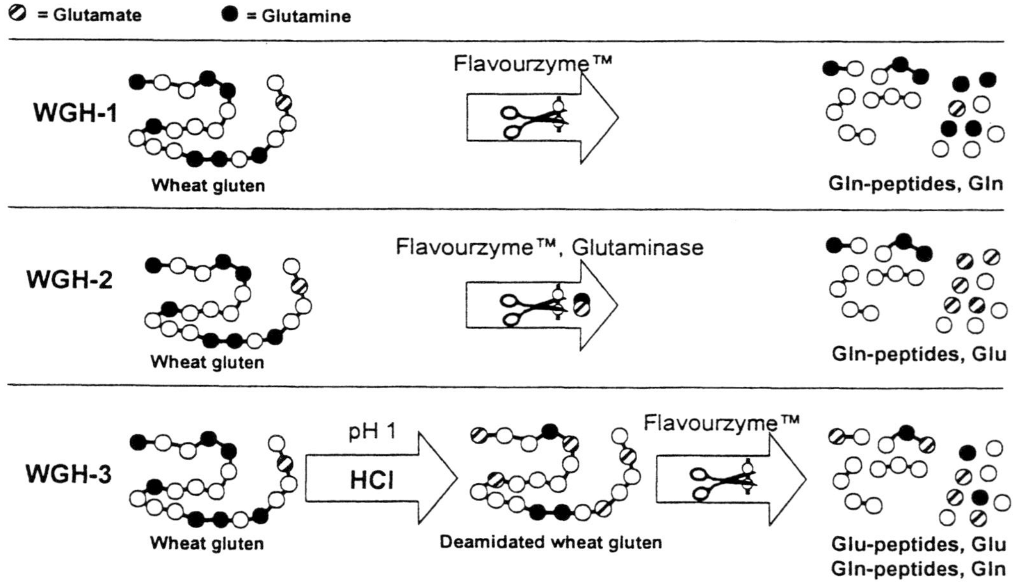

Figure 2. Production of wheat gluten hydrolysates

hydrochloric acid (0.1 mol/L) to partially deamidate protein-bound glutamine into protein-bound glutamic acid. Then, enzymatic hydrolysis with Flavourzyme™ was performed under the same conditions as with the other hydrolysates.

The taste profiles of the wheat gluten hydrolysates (Figure 3) revealed WGH-1 as mainly bitter and only slightly umami. WGH-2 had a considerable umami taste accompanied by bitterness. WGH-3 tasted intensely glutamate-like and was not bitter. Interestingly, the umami intensity of WGH-3 was as high as in WGH-2, although its glutamic acid content was only about one third compared to WGH-2. Therefore, WGH-3 was chosen for further investigation.

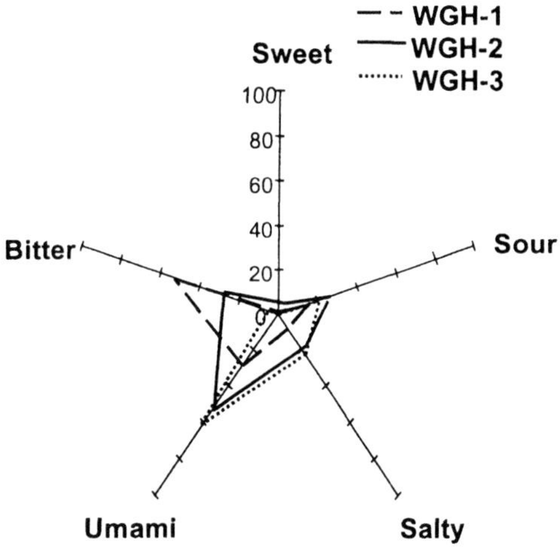

Figure 3. Taste profiles of wheat gluten hydrolysates

The LC-tasting approach was applied to study WGH-3. LC-tasting is the stepwise fractionation by different chromatographic techniques combined with the sensory evaluation of the fractions after removal of the solvents and re-dissolution in water.

Figure 4 shows the LC-tasting procedure employed for WGH-3. At first, an ultrafiltrate of WGH-3 (MW < 3 kDa) was fractionated by size-exclusion chromatography (GPC), which yielded 7 fractions as shown in Figure 5.

The sensory evaluation of the fractions revealed fraction F2 as significantly most intense in umami taste and, in contrast to the other fractions, it was not bitter. Fractions F3 and F4 tasted salty and glutamate-like, but they were less intense and showed a weak bitterness (data not shown).

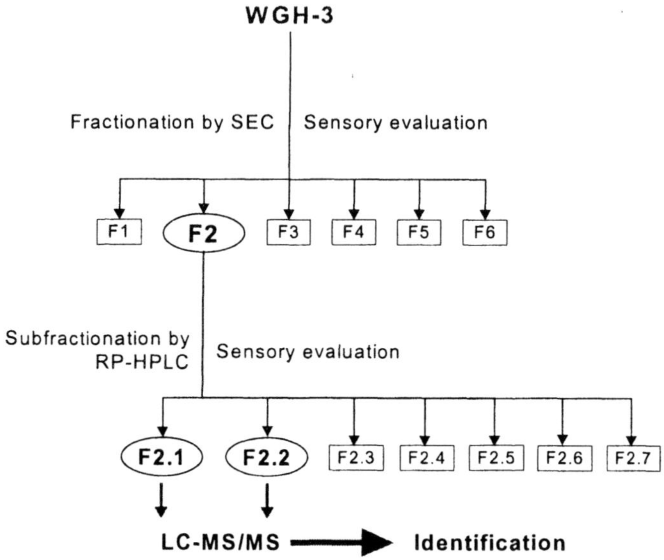

Figure 4. LC-tasting of WGH-3

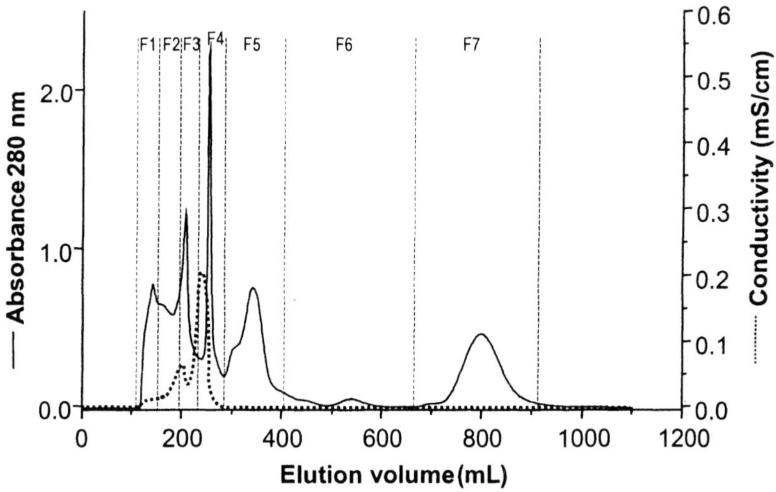

Figure 5. Size exclusion chromatography (GPC) of WGH-3 on Sephadex G10

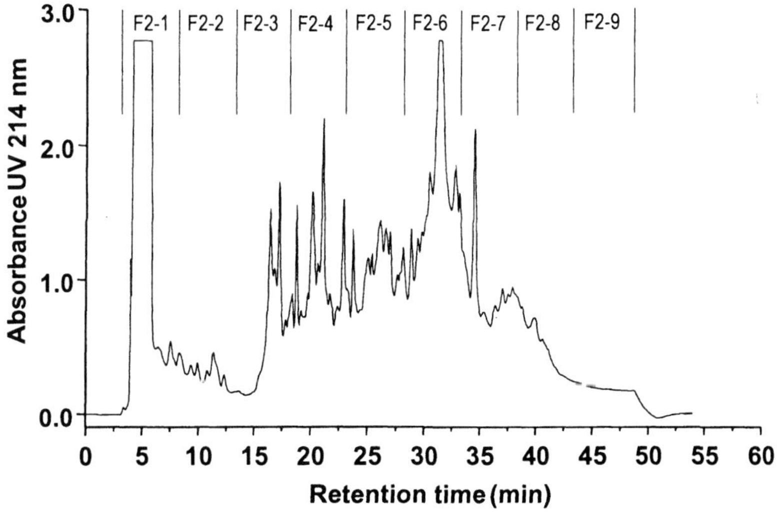

Figure 6. RP-HPLC chromatogram of the taste-active fraction F2 of WGH-3

Table I. Taste Profiles of HPLC Subfractions of WGH-3

attribute	*F2-1*	*F2-2*	*F2-3*	*F2-4*	*F2-5*
sweet	0 ± 0^{b}	0 ± 0^{b}	0 ± 0^{b}	2 ± 1^{c}	0 ± 0^{b}
sour	29 ± 9^{e}	10 ± 6^{d}	14 ± 6^{d}	8 ± 6^{cd}	0 ± 0^{b}
salty	22 ± 14^{c}	3 ± 3^{b}	0 ± 0^{b}	0 ± 0^{b}	0 ± 0^{b}
umami	49 ± 17^{d}	15 ± 5^{c}	10 ± 5^{c}	0 ± 0^{b}	0 ± 0^{b}
bitter	0 ± 0^{b}	15 ± 8^{de}	5 ± 3^{bc}	9 ± 6^{cd}	22 ± 10^{e}
attribute	*F2-6*	*F2-7*	*F2-8*	*F2-9*	
sweet	0 ± 0^{b}	0 ± 0^{b}	0 ± 0^{b}	0 ± 0^{b}	
sour	4 ± 3^{bc}	0 ± 0^{b}	2 ± 1^{b}	0 ± 0^{b}	
salty	0 ± 0^{b}	1 ± 0^{b}	0 ± 0^{b}	0 ± 0^{b}	
umami	0 ± 0^{b}	1 ± 1^{b}	0 ± 0^{b}	0 ± 0^{b}	
bitter	20 ± 10^{e}	31 ± 11^{f}	16 ± 10^{de}	8 ± 4^{bcd}	

[a] Data are expressed as averaged taste intensities of 8 trained panelists scored on a 100 mm line scale (0 = absent; 100 = very strong). [b-f] Mean values within a row showing different letters are significantly different ($p < 0.05$).

The taste-active F2 was subfractionated by RP-HPLC (Figure 6) and the 9 subfractions obtained were assessed sensorially after double freeze-drying to remove the solvents. Table I gives the taste profiles of the nine subfractions.

Fraction F2-1 elicited the most intense umami taste and was slightly sour and salty. Fractions F2-2 and F2-3 tasted also glutamate-like and were accompanied by bitterness. The other subfractions had only weak, mainly bitter taste.

Fraction F2-1 represented the void volume peak in chromatogram which was not separated by RP-HPLC. Separation was achieved by hydrophilic interaction liquid chromatography (HILIC), recently proposed for the separation of peptides and of polar compounds in natural drug discovery (*15, 16*). Interfacing HILIC to electrospray ionization mass spectrometry (ESI-MS) allowed characterization of the compounds (Figure 7). Identification was based on identical retention times and MS-MS spectra with the corresponding reference compounds.

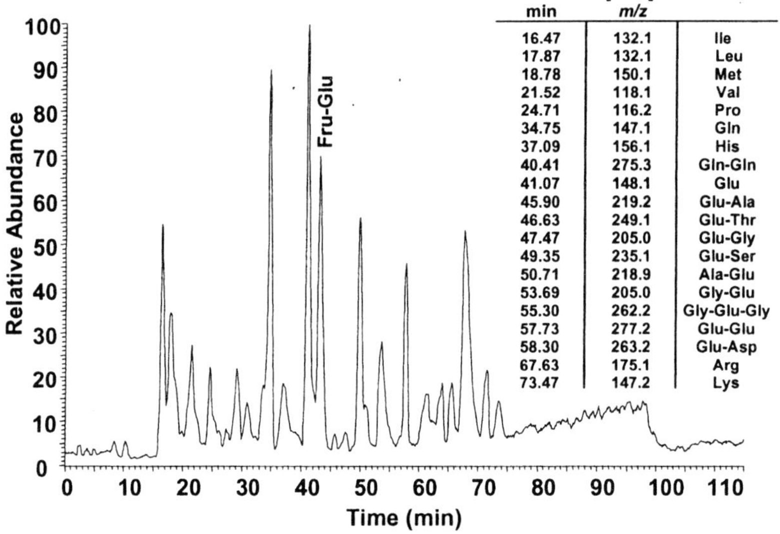

Retention time min	Signal $[M+H]^+$ *m/z*	Compound
16.47	132.1	Ile
17.87	132.1	Leu
18.78	150.1	Met
21.52	118.1	Val
24.71	116.2	Pro
34.75	147.1	Gln
37.09	156.1	His
40.41	275.3	Gln-Gln
41.07	148.1	Glu
45.90	219.2	Glu-Ala
46.63	249.1	Glu-Thr
47.47	205.0	Glu-Gly
49.35	235.1	Glu-Ser
50.71	218.9	Ala-Glu
53.69	205.0	Gly-Glu
55.30	262.2	Gly-Glu-Gly
57.73	277.2	Glu-Glu
58.30	263.2	Glu-Asp
67.63	175.1	Arg
73.47	147.2	Lys

Figure 7. HILIC-ESI base peak chromatogram of the taste-active fraction F2 of WGH-3.

Free amino acids, especially free glutamic acid, glutamine, isoleucine, and leucine, were identified as well as many small glutamyl di- and tripeptides. Several of these glutamyl peptides, in particular Glu-Ser, Ala-Glu, Glu-Asp, and Glu-Glu have been described as eliciting umami and brothy sensory properties

(*7-9, 17*) According to van den Ouweland et al. (*18*) amino acids and peptides individually demonstrate no taste effect, but in combination they generate a general savory taste background. This effect is called "ternary synergism". A similar finding was described by Maehashi et al. (*19*) who identified small hydrophilic glutamyl peptides in chicken protein. Some of the peptides had no umami taste if tested individually, but in combination they elicited an umami taste and enhanced the sensory properties of IMP.

Several signals in the MS chromatogram did not correspond to peptides, but showed neutral losses of 162 and 324 mass units in the MS-MS and MS^3 spectra, which are characteristic for mono- and disaccharide moieties. The remaining residues were assigned to glutamyl glutamate and to the amino acids glutamic acid, glutamine, and lysine. The compounds were identified as Amadori compounds which are glycoconjugates between aldoses and amino acids formed during the early stages of the Maillard reaction. The most abundant Amadori compound detected in the taste-active fraction F2-1 was *N*-(1-deoxy-fructos-1-yl) glutamic acid (Fru-Glu). Figure 8 shows its mass chromatogram and the MS^3 spectrum of *m/z* 292, formed by loss of water from the pseudomolecular parent ion *m/z* 310. The signal at *m/z* 148 corresponds to the glutamic acid fragment and the ion *m/z* 274 results from another loss of a water molecule from the parent ion *m/z* 292.

Fru-Glu is likely to be formed from free glutamic acid present in the hydrolysate and from glucose, which is present in commercial wheat gluten as starch degradation product. Fru-Glu was also found earlier in dried vegetables by Reutter and Eichner (20). Table II shows that the Fru-Glu content is as high as 4 % in drum-dried tomato powder. Higher thermal stress during drying, for example air drying versus freeze-drying, results in higher contents of Fru-Glu, as shown for bell pepper and asparagus powder.

Table II. Contents of *N*-(1-Deoxy-Fructos-1-yl) Glutamic Acid in Dried Vegetables

	Fru-Glu (mg/100 g dry matter)
Leek, air-dried	0
Carrot, air-dried	52
Bell pepper, freeze-dried	58
Bell pepper, air-dried	85
Asparagus, freeze-dried	86
Asparagus, air-dried	126
Cauliflower, air-dried	89
Celery, air-dried	265
Tomato powder, air-dried	2886
Tomato powder, drum-dried	3788

Data taken from reference (*20*)

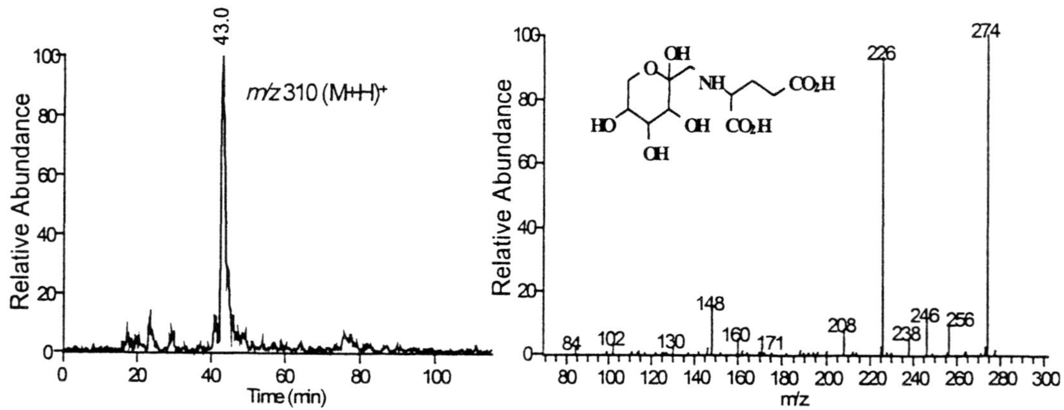

Figure 8. Mass chromatogram and MS³ spectrum of N-(1-deoxy-fructos-1-yl) glutamic acid (positive ionization mode)

Fru-Glu revealed a pronounced umami note in the sensory evaluation at a concentration of 10 mmol/L in water at pH 6.0. Its bouillon-like taste was only slightly weaker compared to a solution of MSG, but it did not elicit the sweetish taste of MSG that is often not desired in culinary products. The same effect was observed when Fru-Glu and MSG were compared in a bouillon base, which did not contain other taste enhancers.

Conclusions

LC-tasting, the combination of chromatographic fractionation methods and sensory evaluation proved to be a useful tool to locate taste-active compounds in a wheat gluten hydrolysate. By employing HILIC-ESI-MS/MS and MS^3 techniques polar glutamyl di- and tripeptides and free amino acids could be separated and characterized in a taste-active subfraction. In addition, the Amadori compound of glutamic acid and glucose was identified as most abundant glycoconjugate present. It elicits a bouillon-like taste similar to MSG, but not its sweetish note. It is concluded that *N*-(1-deoxy-fructos-1-yl) glutamic acid contributes together with small hydrophilic glutamyl peptides and free glutamic acid to the savory taste of WGH-3.

Acknowledgments

We thank Dr. F. Robert for kindly providing a sample of *N*-(1-deoxy-fructos-1-yl) glutamate.

References

1. Yamaguchi S. *J. Food Sci.* **1967**, 32, 473-478.
2. Yamaguchi S., Yoshikawa T., Ikeda S., Ninomiya T. *J. Food Sci.* **1971**, 36, 1761-1765.
3. Velíšek J., Davídek J., Kubelka V., Tran Thi Bich Thu, Hajšlová J. *Nahrung* **1978,** 22, 735-743.
4. Schlichtherle-Cerny H., Grosch W. *Z. Lebensm. Unters. Forsch. A* **1998**, 207, 369-376.
5. Warmke R., Belitz H.D., Grosch W. *Z. Lebensm. Unters. Forsch A* **1996**, 203, 230-235.
6. Monastyrskaia K., Lundstrom K., Plahl D., Acuna G., Schweitzer C., Malherbe P., Mutel V. *Br. J. Pharmacol.* **1999**, 128, 1027-1034.
7. Arai S., Yamashita M., Fujimaki M. *Agric. Biol. Chem.* **1972**, 36, 1253-1256.

8. Arai S., Yamashita M., Fujimaki M.. *Agric. Biol. Chem.* **1973**, 37, 151-156.
9. Noguchi M., Arai S., Yamashita K., Kato H., Fujimaki M. *J. Agric. Food Chem.* **1975**, 23, 49-53.
10. Van den Oord A.H.A., van Wassenaar P.D. *Z. Lebensm. Unters. Forsch. A* **1997**, 205, 125-130.
11. Schmid W., Grosch W. *Z. Lebensm. Unters. Forsch.* **1986**, 182, 407-412.
12. Frank O., Ottinger H., Hofmann T. *J. Agric. Food Chem.* **2001**, 49, 231-238.
13. Schlichtherle-Cerny H., Affolter M., Blank I., Cerny C., Robert F., Beksan E., Hofmann T., Schieberle P. EP 1 252 825A1, **2002**
14. Schlichtherle-Cerny H., Amadò R. *J. Agric. Food Chem.* **2002**, 50, 1515-1522.
15. Yoshida T. *Anal. Chem.* **1997**, 69, 3038-3043.
16. Strege M. A. *Anal. Chem.* **1998**, 70, 2439-2445.
17. Ohyama S., Ishibashi N., Tamura M., Nishizaki H., Okai H. *Agric. Biol. Chem.* **1988**, 52, 871-872.
18. Van den Ouweland G.M.A., Olsman H., Peer H.G. In *Agricultural and Food Chemistry: Past, Present, Future;* Teranishi R., Ed; AVI Publishing Company, Westport, CT, 1978, pp 292-314.
19. Maehashi K., Matsuzaki M., Yamamoto Y., Udaka S. *Biosci. Biotechnol. Biochem.* **1999**, 63, 555-559.
20. Reutter M., Eichner K. *Z. Lebensm. Unters. Forsch.* **1989**, 188, 28-35.

Chapter 15

Application of Bacterial γ-Glutamyl-Transpeptidase to Improve the Taste of Food

Hideyuki Suzuki and Hidehiko Kumagai

Division of Integrated Life Science, Graduate School of Biostudies, Kyoto University, Kyoto 606–8502, Japan

Bacterial γ-glutamyltranspeptidase (GGT, EC 2.3.2.2) was applied to improve the taste of food. We found that the bitterness of amino acids was reduced, sourness produced, and preference increased with γ-glutamylization. An enzymatic method for the synthesis of γ-glutamyl amino acids involving GGT was developed. An enzymatic method involving GGT for the synthesis of theanine, which is the major umami component of tea, was developed. The major umami component of soy sauce is glutamate (Glu), and the effective conversion of Gln to Glu by glutaminase during its fermentation in the presence of 18% NaCl is a matter of concern. Purified *Bacillus subtilis* GGT was salt-tolerant and 76% of its original glutaminase activity remained even in the presence of 18% NaCl.

GGT is the key enzyme in glutathione metabolism and is widely distributed in living things (*1-4*). GGT consists of one large subunit and one small subunit. The molecular weight of the large subunit is about 40.000 and that of the small subunit is about 20.000. It was known that the large and small subunits of mature GGT are generated from a common enzymatically inactive precursor through post-translational proteolytic processing (*5-13*). We recently showed that this processing is an autocatalytic event and that the catalytic nucleophile for the processing is the oxygen atom of the side-chain of the Thr residue that is going to be the N-terminal of the small subunit after processing (*14*). This oxygen atom is also the nucleophile for the enzymatic reaction (*15*). GGT catalyzes two reactions (Figure 1). The reactions catalyzed by GGT proceed via a γ-glutamyl-enzyme intermediate. If this intermediate undergoes nucleophilic substitution by amino acids or peptides, it is the transpeptidation reaction yielding new γ-glutamyl compounds. However, if the intermediate undergoes nucleophilic substitution by water, it is the hydrolysis reaction, which releases glutamic acid. When the original γ-glutamyl compound is glutamine, the hydrolysis reaction is a "glutaminase" reaction. The pH optima of the two reactions are different. Therefore, by adjusting the reaction pH, we can make the enzyme catalyze the transpeptidation reaction selectively. Employing various γ-glutamyl acceptors, we can synthesize various γ-glutamyl compounds using GGT.

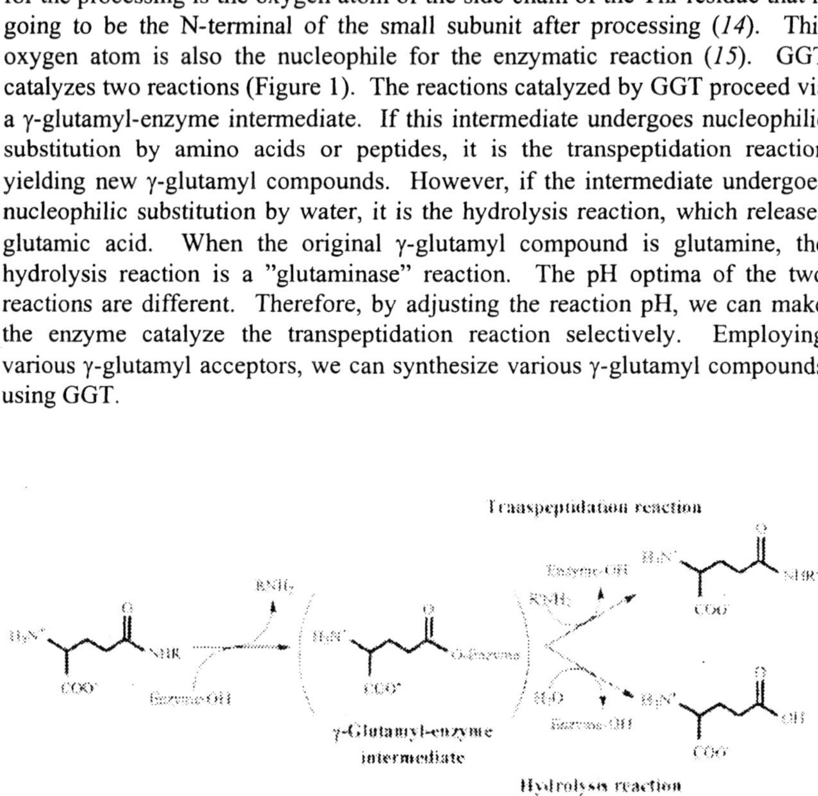

Figure 1. The mechanisms of the enzymatic reactions catalyzed by GGT.

Why are we interested in γ-glutamyl compounds? One reason is that their solubility in water is increased by γ-glutamylization. Another reason is that the γ-glutamyl linkage is resistant to attack by peptidases in serum. Therefore, γ-glutamyl compounds can possibly be used as pro-drugs for specific organs that express GGT, as in the cases of γ-Glu-DOPA (*16-19*) and γ-Glu-dermorphin (*20*). Another reason is that some γ-glutamyl compounds taste good. In this chapter, we describe the application of bacterial GGT to improve the taste of food using both its transpeptidation and hydrolysis activities.

Utilization of the Transpeptidation Reaction of GGT to Improve the Taste of Food

Glutathione, which is one of the major γ-glutamyl compounds in living things, tastes sour, but has a preferable and refreshing lemon-like sourness. We found that the taste of γ-Glu-Phe was strikingly different from that of Phe. γ-Glu-Phe tasted just like glutathione.

Effect of γ-Glutamylization of Bitter Amino Acids

Some L-amino acids taste bitter. Aromatic amino acids, basic amino acids, and branched-chain amino acids taste bitter. Unfortunately, many bitter amino acids are essential ones. Therefore, when we take an amino acid mixture orally, the bitterness of these amino acids is a crucial problem. These days amino acid mixtures consisting of branched-chain amino acids, Phe, Arg, and/or Lys, are very popular in Japan as supplements. However, they taste extremely bitter, and thus are hard to take orally without sweetening and flavoring.

Phe is an essential amino acid and one of the most bitter ones. First, the taste of Phe and γ-Glu-Phe was evaluated by eight panel members. As shown in Figure 2, various members evaluated Phe to be bitter at 15 mM. However, γ-Glu-Phe did not show any bitterness at all. Phe was practically not sour, but the sourness of γ-Glu-Phe was obvious. Also the preference for γ-Glu-Phe was better than that for Phe. Therefore, it was evident that γ-glutamylization of Phe improved its taste. Several panel members said γ-Glu-Phe had a refreshing lemon-like sourness, and we think this was the reason why they preferred γ-Glu-Phe. When an equimolar amount of glutamic acid was added to Phe, the mixture was sourer than Phe alone, however, there was no change in the bitterness or preference. This indicates that the γ-glutamyl linkage is necessary for the effect and that the existence of Glu is not enough. The sourness of a mixture of Phe and Glu was not refreshing. Moreover, when Glu was added, the taste became harsh. This might be the reason why the addition of Glu was not effective. There was no significant difference in taste between Phe and γ-Glu-Phe neutralized with NaOH. This suggests that the sourness of γ-Glu-Phe is critical. The effect of γ-glutamylization on the taste of other bitter amino acids was also examined, and similar results were obtained for Val, Leu, and His (Figure 3).

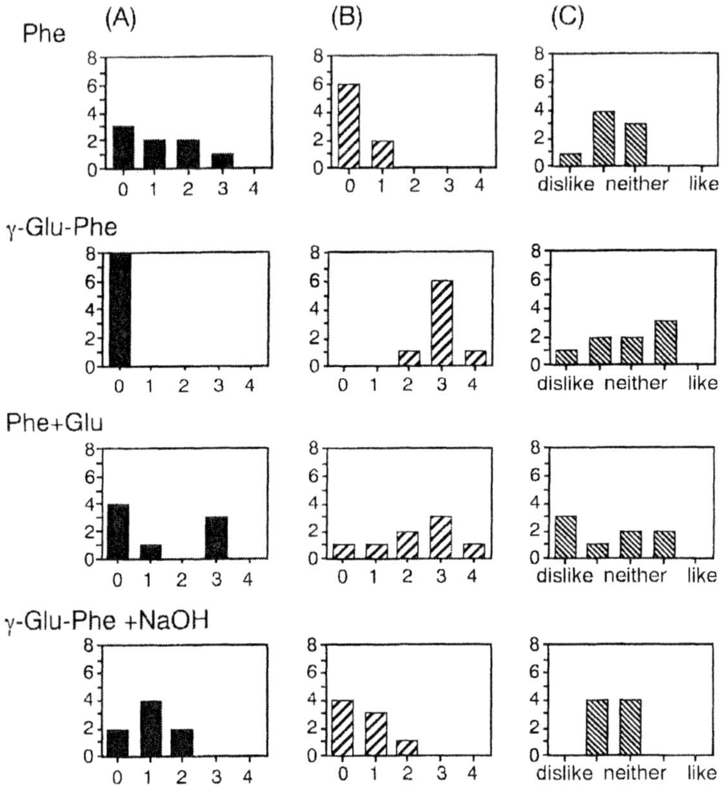

Figure 2. Comparison of the taste of Phe and γ-Glu-Phe at 15 mM. (A) Bitterness, (B) sourness, and (C) preference. The vertical axes indicate the number of people. The horizontal axes in (A) and (B) indicate the intensity of the taste: 0, did not feel; 1, felt slightly; 2, felt weakly; 3, felt; 4, felt strongly. The horizontal axes in (C) indicate the preference on a 5-point scale. (Adapted with permission from reference 21. Copyright 2002 American Chemical Society.)

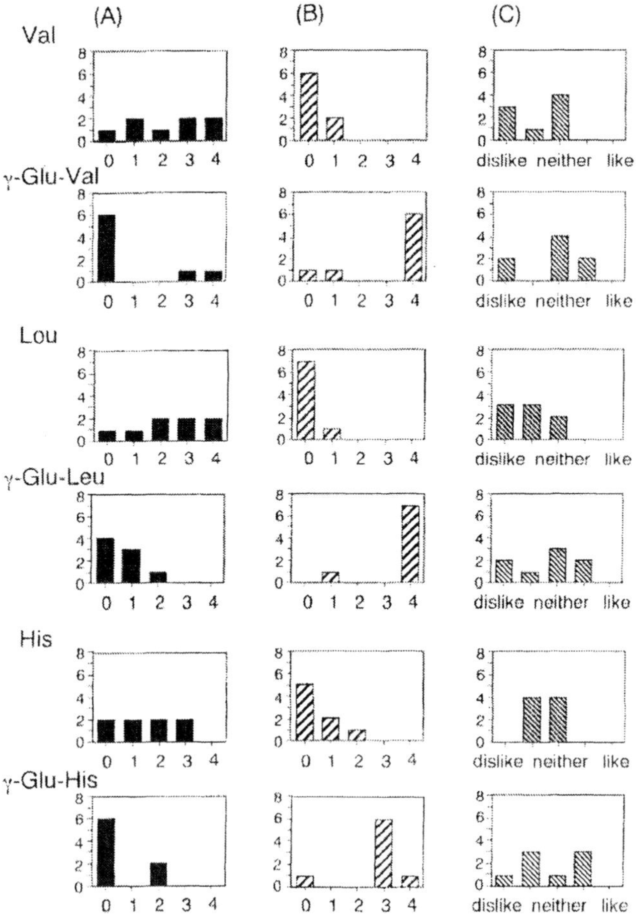

Figure 3. Effect of γ-glutamylization of Val, Leu, and His. (A) Bitterness, (B) sourness, and (C) preference. The axes indicate the same as in Figure 2. The concentrations used were 50 mM for Val and Leu, and their γ-glutamyl derivatives, and 70 mM for His and γ-Glu-His. (Reproduced with permission from reference 21. Copyright 2002 American Chemical Society.)

Kirimura *et al.* (*22*) compared the taste of various dipeptides, and reported that α-Glu-Phe was bitter and sour, while γ-Glu-Phe was not bitter but sour. However, they also reported that all γ-glutamyl amino acids they tested were astringent. As far as we examined, γ-Glu-Phe, γ-Glu-Val, γ-Glu-Leu, and γ-Glu-His were not astringent at all, which is inconsistent with their results.

In conclusion, γ-glutamylization reduced the bitterness, and increased the sourness of and preference for bitter amino acids. γ-Glutamylization should be a powerful method for improving the taste of bitter amino acids.

The question may arise as to how γ-glutamyl amino acids are taken up in the intestines of humans and then metabolized. One possibility is that the γ-glutamyl linkage is cleaved by highly abundant GGT on the brush borders of the columnar epithelial cells on the tips of the villi of the small intestine (*23, 24*), and the released Glu and amino acids are taken up through regular transport systems for amino acids. This is similar to the mechanism of glutathione transport that Inoue suggested (*25*). Another possibility is that γ-glutamyl amino acids are taken up intact and the γ-glutamyl linkage is cleaved by GGT existing in the kidneys.

Enzymatic Production of γ-Glutamyl Amino Acids Using Bacterial GGT

Among the bitter amino acids we tested, the effect of γ-glutamylization was most remarkable for Phe. Therefore, the reaction conditions for the synthesis of γ-Glu-Phe involving bacterial GGT were optimized recently (*21*). Besides, we have already developed an enzymatic method for synthesizing various γ-glutamyl amino acids involving bacterial GGT as a catalyst (*26-32*). Bacterial GGTs are superior to eukaryotic GGTs, because bacterial GGTs are either periplasmic (*33*) or extracellular enzymes (*34*), and can be purified as soluble enzyme preparations from tank cultures, whereas eukaryotic GGTs are membrane-bound enzymes (*1, 2*). The characteristics of our method can be summarized as follows:

- A less expensive γ-glutamyl donor, L-glutamine, can be used effectively as well as glutathione, unlike in the case of mammalian GGT.
- Since GGT is a transferase and not a synthetase, it does not require any energy source such as ATP.
- The substrate specificity of GGT for γ-glutamyl acceptors is quite broad and thus various γ-glutamyl compounds can be synthesized.
- The transpeptidation reaction can occur at various pHs.
- A lot of GGT is readily available because both *E. coli* and *B. subtilis* GGTs

can be purified by simple two-step purification procedures from over-expressing strains (*34-36*).

- Unlike a chemical synthetic method, the protection and deblocking of reactive groups of the substrates are not required.

The optimum reaction conditions and yields of various γ-glutamyl compounds synthesized with our method are summarized in Table I.

Table I. The Reaction Conditions and Yields of Various γ-Glutamyl Amino Acids Synthesized with Our Method

γ-Glutamyl Amino Acid Synthesized	*Concentration*			*pH*	*Yield*	
	L Gln (mM)	*Acceptor (mM)*	*GGT (mU/ml)*		*(%)*	*(g/L)*
γ-Glu-L-DOPA[a]	200	200	250	10.6	79	51.5
γ-Glu-L-His	300	300	200	9.7	48	41.2
γ-Glu-L-Tyr-methylester	300	300	200	7.3	37	35.7
S-Benzyl-glutathione monomethylester	200	100	200	6.2	76	31.2
γ-Glu-L-Phe	200	200	500	10.4	70	41.2
γ-Glu-taurine	200	200	200	10.0	23	11.4

NOTE: *a*, DOPA stands for 3,4-dihydroxyphenylalanine.

SOURCE: Adapted with permission from reference 4. Copyright 1999 Elsevier.

Enzymatic Production of Theanine, An "Umami" Component of Tea Using Bacterial GGT

Theanine (γ-glutamylethylamide), which was first identified by Sakato (*37*) in tea leaves, is not only the main free amino acid component of tea (more than half), but also the major umami component of tea. Ekborg-Ott *et al.* found that tea contains D-theanine as well as L-theanine, and that both of them have similar umami and sweet taste (*38*). There is a positive correlation between the grade of Japanese green tea and the content of theanine (*39, 40*). Theanine is synthesized from Glu and ethylamine by theanine synthetase [L-glutamate:ethylamine ligase, EC 6.3.1.6] using ATP (*41*) in the roots of tea trees (*Camellia sinensis*), and thus synthesized theanine is transported to the leaves and accumulated there (*42, 43*). On exposure to the sunshine, theanine is converted to catechin, which has an astringent taste (*44, 45*). Therefore, tea trees whose leaves are to become "Maccha" and "Gyokuro", green teas of the highest grade, are shaded by screens to block the sunshine after budding. And about 20 days after budding, the two leaves and a bud at the top of each shoot, which are the first crop of the year, are picked (http://www.kyocha.or.jp/english/study). However, even green tea of the highest grade contains on average only 1.5-2% of theanine.

L-Theanine is known to have favorable physiological effects on mammals.

- Decrease in blood pressure of spontaneously hypertensive rats (*46*).
- Inhibitory effects on the convulsive action and spontaneous activity caused by caffeine administration to mice (*47, 48*).
- Inhibitory effect on the excitation caused by caffeine in rats (*49*).
- A feeling of relaxation in humans (*50*).
- Prevention of ischemic neuronal death in gerbils (*51*).

It has been suggested that orally administered theanine is absorbed intact through the intestinal tract into the blood circulation (*52, 53*). Also, theanine is known to be incorporated into the brain through the blood-brain barrier via a Leu-preferring transport system, and it may affect the metabolism and/or the release of some neurotransmitters in the brain (*54*). Therefore, theanine could be a new food additive with specific functions, and several investigators have studied its effective production. A chemical synthetic method has been reported, but it requires protection and deblocking of reactive groups (*55*). Tachiki *et al.* (*56*) reported the production of theanine from Glu and ethylamine through the coupling of baker's yeast preparations and bacterial glutamine synthetase. Because of the low reactivity of glutamine synthetase with ethylamine and the difficulty in pH control of the reaction mixture, the yield was not satisfactory. The production of theanine with cultured cells of *C. sinensis* has been performed, however, not only did it take four weeks for cultivation, but also the yield was low (*57, 58*). Enzymatic production of theanine with glutaminase from *Pseudomonas nitroreducens* has been reported (*59*), however, methods for the overproduction and simple purification of this enzyme have not yet been established. Abelian *et al.* employed immobilized *P. nitroreducens* cells to overcome these drawbacks (*60, 61*).

Regardless of the source, it was not known if GGT can utilize an alkylamine, such as ethylamine, as a γ-glutamyl acceptor for the transpeptidation reaction. Having optimized the reaction conditions, we developed an enzymatic method for synthesizing L-theanine in a relatively high yield (61% against Gln) (*32*) (Figure 4).

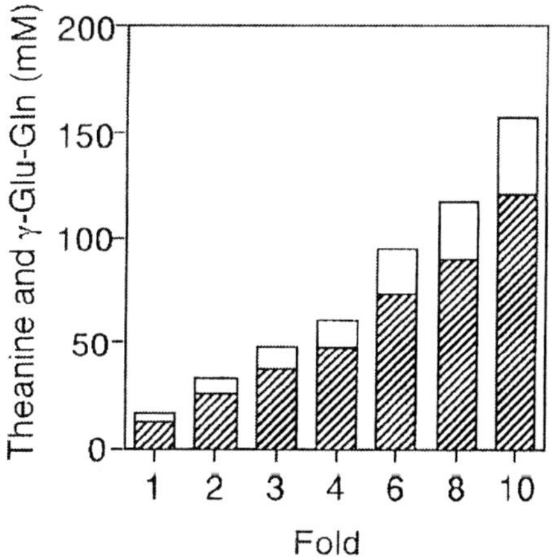

Figure 4. Effects of increasing concentrations of the substrates and GGT on theanine synthesis. The 1-fold reaction mixture contained 20 mM Gln, 150 mM ethylamine and 0.04 units/ml GGT, the 2-fold one 40 mM Gln, 300 mM ethylamine and 0.08 units/ml GGT, and so forth. The reaction was carried out at 37°C for 5 h. The concentration of theanine is shown as slashed boxes and that of γ-Glu-Gln as open boxes. (Reproduced with permission from reference 32. Copyright 2002 Elsevier.)

Utilization of the Hydrolysis Reaction of GGT to Improve The Taste of Food

As described above, if the substrate for the hydrolysis reaction of GGT is glutamine, the reaction is a "glutaminase" reaction.

Soy sauce is a traditional Japanese seasoning and is also an international one today. Its delicious taste mainly depends on the amount of glutamic acid. During its fermentation, soy proteins are digested into peptides by proteases from *Aspergillus oryzae* or *A. sojae*, and the peptides are then cleaved into amino acids by their peptidases. The glutamine liberated is hydrolyzed to glutamic acid by glutaminase (Figure 5). When the level of glutaminase is insufficient, glutamine is converted spontaneously to tasteless or slightly sour pyroglutamic acid. Therefore, glutaminase is one of the most important enzymes for flavor enhancement in the manufacture of soy sauce. Soy sauce fermentation is performed in the presence of 18% NaCl at pH 5.5 to prevent contamination. Eighteen % NaCl corresponds to more than 3 M. In the presence of this much NaCl, the activity of *Aspergillus* glutaminases is strongly inhibited. Therefore, salt-tolerant glutaminases were searched for in microorganisms for application to soy sauce fermentation.

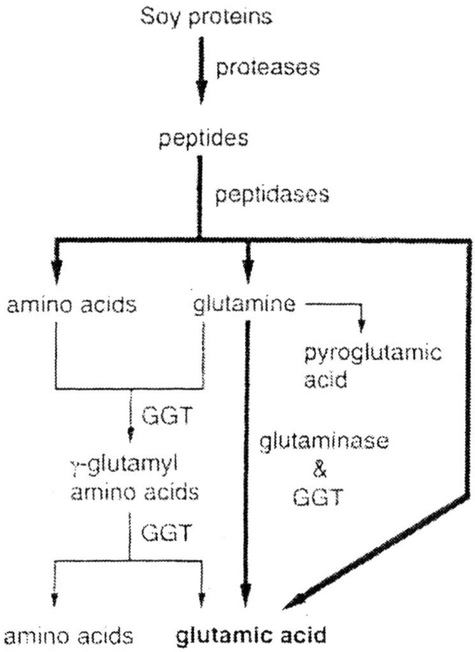

Figure 5. Digestion of soy proteins during the fermentation of soy sauce.

We found that some *Bacillus* species synthesize salt-tolerant GGTs (*34*). We decided to use *B. subtilis* 168 because the DNA sequence of its genome is already known and genetic methods are well developed for this strain. The *spoOA abrB* mutant (strain SWV150) was obtained from Dr. Strauch, and his research group has already reported that the expression of GGT is regulated by these regulators (SpoOA and AbrB) in *B. subtilis* (*62*). Its *ggt* gene was cloned on a plasmid vector (pHY300PLK) to obtain pMH2312, and strain SWV150 was transformed with this plasmid to obtain strain MH2308. GGT was overproduced in strain MH2308 (Figure 6). GGT was purified from the culture medium and its salt-tolerance was checked using the purified enzyme. As

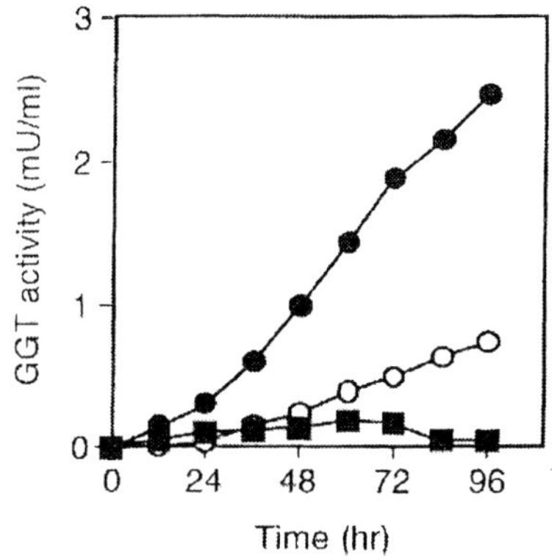

Figure 6. GGT activity during bacterial growth. The graph shows the parental strain (□), SWV150 (○), and MH2308 (●). (Reproduced with permission from reference 34. Copyright 2003 Elsevier.)

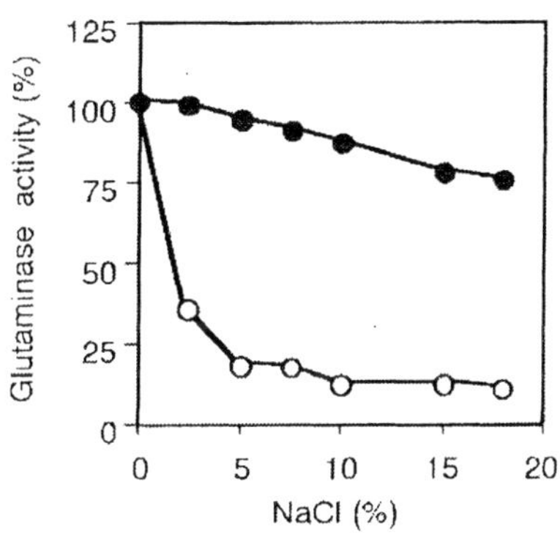

Figure 7. Effects of the NaCl concentration on the glutaminase activities of B. subtilis (○) and E. coli (●) GGT. The reaction mixtures contained 0-18% NaCl. The respective concentrations of NaCl were added to the reaction mixtures on preincubation. Glutaminase activity was measured using L-glutamine as a substrate. (Reproduced with permission from reference 34. Copyright 2003 Elsevier.)

shown in Figure 7, *B. subtilis* GGT was quite salt-tolerant, and even in the presence of 18% NaCl, 76% of its original glutaminase activity remained.

Conclusion

We have shown that both the transpeptidation and hydrolysis reactions of GGT are very useful for improving the taste of food. In conclusion, bacterial GGT is a super enzyme for the food industry. The use of bacterial GGT should lead us to a tasty life.

Acknowledgments

We would like to thank Yoko Kajimoto, Nobukazu Miyakawa, Hiromichi Minami, and Shunsuke Izuka of our laboratory for their excellent assistance. This work was supported by Grants-in-Aid for Scientific Research, No. 13660090 to H.S. and No. 14360056 to H.K., from the Ministry of Education, Culture, Sports, Science, and Technology of Japan, and by research funds to H.S. from the Society for Umami Research, the Asahi Brewery Foundation, and the Fuji Foundation for Protein Research.

References

1. Tate, S. S.; Meister, A. *Mol. Cell. Biochem.* **1981,** *39,* 357-368.
2. Taniguchi, N.; Ikeda, Y. *Adv. Enzymol. Rel. Areas Mol. Biol.* **1998,** *72,* 239-278.
3. Suzuki, H.; Hashimoto, W.; Kumagai, H. *J. Bacteriol.* **1993,** *175,* 6038-6040.
4. Suzuki, H.; Hashimoto, W.; Kumagai, H. *J. Mol. Catal.* **1999,** *B 6,* 175-184.
5. Nash, B.; Tate, S. *J. Biol. Chem.* **1982,** *257,* 585-588.
6. Capraro, M. A.; Hughey, R. *FEBS Lett.* **1983,** *157,* 139-143.
7. Matsuda, Y.; Tsuji, A.; Kuno, T.; Katunuma, N. *J. Biochem. (Tokyo)* **1983,** *94,* 755-765.
8. Kuno, T.; Matsuda, Y.; Katunuma, N. *Biochem. Biophys. Res. Commun.* **1983,** *114,* 889-895.
9. Yokosawa, N.; Taniguchi, N.; Tsukada, Y.; Makita, A. *Oncodev. Biol. Med.* **1983,** *4,* C71-78.
10. Nash, B.; Tate, S. *J. Biol. Chem.* **1984,** *259,* 678-685.
11. Finidori, J.; Laperche, Y.; Haguenauer-Tsapis, R.; Barouki, R.; Guellaen, G.; Hanoune, J. *J. Biol. Chem.* **1984,** *259,* 4687-4690.
12. Barouki, R.; Finidori, J.; Chobert, M. N.; Aggerbeck, M.; Laperche, Y.; Hanoune, J. *J. Biol. Chem.* **1984,** *259,* 7970-7974.

13. Hashimoto, W.; Suzuki, H.; Nohara, S.; Kumagai, H. *Biochem. Biophys. Res. Commun.* **1992,** *189,* 173-178.
14. Suzuki, H.; Kumagai, H. *J. Biol. Chem.* **2002,** *277,*43536-43543.
15. Inoue, M.; Hiratake, J.; Suzuki, H.; Kumagai, H.; Sakata, K. *Biochemistry* **2000,** *39,* 7764-7771.
16. Wilk, S.; Mizoguchi, H.; Orlowski, M. *J. Pharmacol. Exp. Ther.* **1978,** *206,* 227-232.
17. Worth, D. P.; Harvey, J. N.; Brown, J.; Lee, M. R. *Clin. Sci.* **1985,** *69,* 207-214.
18. Ichinose, H.; Togari, A.; Suzuki, H.; Kumagai, H.; Nagatsu, T. *J. Neurochem.* **1987,** *49,* 928-932.
19. Boateng, Y. A.; Barber, H. E.; MacDonald, T. M.; Petrie, J. C.; Lee, M. C; Whiting, P. H. *Br. J. Pharmacol.* **1990,** *101,* 301-306.
20. Misicka, A.; Maszczynska, I.; Lipkowski, A. W.; Stropova, D.; Yamamura, H. I.; Hruby, V. J. *Life Science* **1996,** *58,* 905-911.
21. Suzuki, H.; Kajimoto, Y.; Kumagai, H. *J. Agric. Food Chem.* **2002,** *50,* 313-318.
22. Kirimura, J.; Shimizu, A.; Kimizuka, A.; Ninomiya, T.; Katsuya, N. *J. Agric. Food Chem.* **1969**, *17,* 689-695.
23. Ross, L. L.; Barber, L.; Tate, S. S; Meister, A. *Proc. Natl. Acad. Sci. USA* **1973**, *70,* 2211-2214.
24. Nakamura, Y.; Azuma, T.; Fukuyama, H.; Suzuki, F.; Nagata, Y. *J. Nutr. Sci. Vitaminol.* **1985**, *31,* 179-187.
25. Inoue, M. In *Renal Biochemistry*; Kinne, R. K. H., Ed.; Elsevier: Amsterdam, The Netherlands, 1985; pp 225-269.
26. Kumagai, H.; Echigo, T.; Suzuki, H.; Tochikura, T. *Agric. Biol. Chem.* **1988**, *52,* 1741-1745.
27. Kumagai, H.; Suzuki, H.; Shimizu, M.; Tochikura, T. *J. Biotechnol.* **1989,** *9,* 129-138.
28. Kumagai, H.; Echigo, T.; Suzuki, H.; Tochikura, T. *Agric. Biol. Chem.* **1989,** *53,* 1429-1430.
29. Kumagai, H.; Echigo, T.; Suzuki, H.; Tochikura, T. *Lett. Appl. Microbiol.* **1989,** *8,* 143-146.
30. Kumagai, H.; Suzuki, H.; Echigo, T.; Tochikura, T. *Ann. N.Y. Acad. Sci.* **1990,** *613,* 647-651.
31. Suzuki, H.; Miyakawa, N.; Kumagai, H. *Enzyme Microb. Technol.* **2002,** *30,* 883-888.
32. Suzuki, H.; Izuka, S.; Miyakawa, N.; Kumagai, H. *Enzyme Microb. Technol.* **2002,** *31,* 884-889.
33. Suzuki, H.; Kumagai, H.; Tochikura, T. *J. Bacteriol.* **1986,** *168,* 1332-1335.

34. Minami, H.; Suzuki, H.; Kumagai, H. *Enzyme Microb. Technol.* **2003,** *32,* 431-438.
35. Suzuki, H.; Kumagai, H.; Echigo, T.; Tochikura, T. *Biochem. Biophys. Res. Commun.* **1988,** *150,* 33-38.
36. Claudio, J. O.; Suzuki, H.; Kumagai, H.; Tochikura, T. *J. Ferment. Bioeng.* **1991,** *72,* 125-127.
37. Sakato Y. *Nippon Nogeikagaku Kaishi (in Japanese)* **1949,** *23,* 262-267.
38. Ekborg-Ott, K. H.; Taylor, A.; Armstrong, D. W. *J. Agric. Food Chem.* **1997,** *45,* 353-363.
39. Nakagawa, M. *Jpn. Agric. Res. Q.* **1970,** *5,* 43-47.
40. Goto, T.; Yoshida, Y.; Amano, I.; Horie, H. *Foods Food Ingredients J. Jpn.* **1996,** *170,* 46-51.
41. Sasaoka, K.; Kito, M.; Onishi, Y. *Agric. Biol. Chem.* **1965,** *29,* 984-988.
42. Konishi, S.; Takahashi, E. *Nippon Dojo-Hiryogaku Zasshi (in Japanese)* **1969,** *40,* 479-484.
43. Wickremasinghe, R. L.; Perera, K. P. W. C. *Tea Q.* **1972,** *43,* 175-179.
44. Kito, M.; Kokura, H.; Izaki, J.; Sasaoka, K. *Agric. Biol. Chem.* **1966,** *30,* 623-624.
45. Kito, M.; Kokura, H.; Izaki, J.; Sasaoka, K. *Phytochemistry* **1968,** *7,* 599-603.
46. Yokogoshi, H.; Kato, Y.; Sagesaka, Y. M.; Takihara-Matsuura, T.; Kakuda, T.; Takeuchi, N. *Biosci. Biotechnol. Biochem.* **1995,** *59,* 615-618.
47. Kimura, R.; Murata, T. *Chem. Pharm. Bull.* **1971,** *19,* 1257-1261.
48. Kimura, R.; Kurita, M.; Murata, T. *Yakugaku Zasshi (in Japanese)* **1975,** *95,* 892-895.
49. Kakuda, T.; Nozawa, A.; Unno, T.; Okamura, N.; Okai, O. *Biosci. Biotechnol. Biochem.* **2000,** *64,* 287-293.
50. Kobayashi, K.; Nagato, Y.; Aoi, N.; Juneja, L. R.; Kim, M.; Yamamoto, T.; Sugimoto, S. *Nippon Nogeikagaku Kaishi (in Japanese)* **1998,** *72,* 153-157.
51. Kakuda, T.; Yanase, H.; Utsunomiya, K.; Nozawa, A.; Unno, T.; Kataoka, K. *Neurosci. Lett.* **2000,** *289,* 189-192.
52. Kitaoka, S.; Hayashi, H.; Yokogoshi, H.; Suzuki, Y. *Biosci. Biotechnol. Biochem.* **1996,** *60,* 1768-1771.
53. Unno, T.; Suzuki, Y.; Kakuda, T.; Hayakawa, T.; Tsuge, H. *J. Agric. Food Chem.* **1999,** *47,* 1593-1596.
54. Yokogoshi, H.; Kobayashi, M.; Mochizuki, M.; Terashima, T. *Neurochem. Res.* **1998,** *23,* 667-673.
55. Sakato, Y.; Hashizume, T.; Kishimoto, Y. *Nippon Nogeikagaku Kaishi (in Japanese)* **1949,** *23,* 269-271.
56. Tachiki, T.; Suzuki, H.; Wakisaka, S.; Yano, T.; Tochikura, T. *J. Gen. Appl. Microbiol.* **1986,** *32,* 545-548.

57. Matsuura, T.; Kakuda, T. *Agric. Biol. Chem.* **1990,** *54,* 2283-2286.
58. Orihara, T.; Furuya, T. *Plant Cell Reports* **1990,** *9,* 65-68.
59. Tachiki, T.; Yamada, T.; Mizuno, K.; Ueda, M.; Shiode, J.; Fukami, H. *Biosci. Biotechnol. Biochem.* **1998,** *62,* 1279-1283.
60. Abelian, V. H.; Okubo, T.; Shamtsian, M. M.; Mutoh, K.; Chu, D.-C.; Kim, M.; Yamamoto, T. *Biosci. Biotechnol. Biochem.* **1993,** *57,* 481-483.
61. Abelian, V. H.; Okubo, T.; Mutoh, K.; Chu, D.-C.; Kim, M.; Yamamoto, T. *J. Ferment. Bioeng.* **1993,** *76,* 195-198.
62. Xu, K.; Strauch, M. A. *J. Bacteriol.* **1996,** *178,* 4319-4322.

Flavor Interactions

Chapter 16

Effects of Viscosity on Flavor Perception: A Multimodal Approach

David J. Cook[1], Tracey A. Hollowood[1], Elodie Pettelot[2], and Andrew J. Taylor[1]

[1]Samworth Flavor Laboratory, Division of Food Sciences, The University of Nottingham, Sutton Bonington Campus, Loughborough, Leicestershire LE12, 5RD, United Kingdom
[2]ENSBANA, 1 Esplanade Erasme, 21 000 Dijon, France

Many compounds that increase the viscosity of liquid foods also have an impact on taste and aroma perception. Whilst specific effects might result from aroma binding to macromolecules, the fundamental changes in perception arising from presentation of viscous stimuli are of more generic interest. This paper examines the poorly understood phenomenon of sweetness and aroma suppression in viscous hydrocolloid solutions. Possible physical causes are discussed alongside a multi-sensory approach to flavor perception developed to account for observed perceptual changes. Oral perception of viscosity can be correlated with the shear-stress developed in-mouth when manipulating liquid samples. Sensed by mechano-receptors, this forms part of the somatosensory input for the oral cavity. It is hypothesized that sensory signals relating to the mouthfeel of a fluid can interact at both neural and psychological levels with the perception of taste and aroma.

It is well documented that as the viscosity of a liquid food is increased, there are changes in flavor perception. Generally, flavor intensity decreases as viscosity is increased, although specific effects on flavor perception may depend upon:

- The nature and concentration of the viscosifying agent
- The taste qualities and aromas under consideration and their concentrations relative to sensory thresholds
- Physicochemical interactions between flavor molecules and the thickener (e.g. binding, ion-exchange or salting in / out effects)
- Taste effects of the thickener itself
- The sensory methodology used to investigate changes

Due to this wide range of variables, even in a model system, it is often difficult to predict the sensory effects of increasing viscosity. However, one feature of increasing viscosity appears to be more fundamental: it causes a decrease in sweetness perception (*1*).

Many workers have used hydrocolloid thickeners to manipulate viscosity in model systems incorporating taste and aroma compounds. Baines & Morris (*2*) investigated sensory perception of sweetness and of strawberry flavor in solutions thickened with guar gum. Perception of both attributes was unaffected at low thickener concentrations, relative to that in water. However, when the thickener concentration increased above c* (the coil-overlap concentration; the point at which there is an abrupt increase in viscosity as thickener is added to the system) sensory perception of both sweetness and strawberry flavor were progressively suppressed. The viscosity of most hydrocolloids at the c* concentration in water is approximately 10mPas. It was postulated that inefficient mixing in solutions above c*, caused by the development of an entangled polymer network, inhibited the transport of taste and aroma molecules to their respective receptors. With the advent of techniques such as API-MS, it has become possible to measure the in-nose release of aroma compounds on a breath by breath basis and thus to test this hypothesis. Hollowood *et al.* (*3*) investigated sensory perception and in-nose aroma release from solutions sweetened with sucrose, flavored with benzaldehyde and thickened with hydroxypropylmethylcellulose (HPMC). In agreement with previous findings, both sweetness and almond flavor perception were suppressed at thickener concentrations above c*. However, aroma delivery in-nose whilst drinking the solutions was shown to be independent of thickener concentration. It was concluded that a drop in perceived sweetness was driving the fall in perception of a congruent sweet aroma (due to a taste-aroma interaction (*4*)) even though the aroma stimulus in-nose remained constant.

The perception of viscosity

When liquid foods are placed in the oral cavity, we are able to sense a physical presence and to appraise a range of textural qualities, including viscosity. This is achieved by applying a shear to the fluid, manipulating it between the tongue and the roof of the mouth, and sensing how the material flows. Tactile stimuli in the mouth are sensed by mechanoreceptors. This input forms part of the somatic sensory system, which also functions to detect pain (nociception) and temperature.

Many workers have tried to correlate rheological measurements with the sensory perception of viscosity (*5-7*). Since hydrocolloids are non-Newtonian in nature, their apparent viscosity varies with applied shear. There has been much debate over the shear rates that liquid samples are subjected to in-mouth and hence the appropriate viscosity measurement with which to correlate. In a refinement of this approach, Jozef Kokini (*8*) used a rheological model to calculate the shear stress developed on the tongue as a result of manipulating non-Newtonian fluids in-mouth. Analysis of the forces generated gave an equation whose principal variables were the power law parameters of the particular fluid. These define the rate at which a sample thins under increasing rates of shear and can be calculated from simple rheological measurements (*9*). The validity of this equation has been demonstrated in experiments showing a direct correlation between this mathematical shear stress and the sensory perception of 'thickness' (*8,9*).

Texture-taste interactions

Cook *et al.* (*10*) recently demonstrated that sweetness and banana flavor perception in hydrocolloid thickened solutions could be modeled in terms of the Kokini oral shear stress, the physical stimulus for the sensory perception of viscosity. Models were sufficiently robust to explain the behaviors of five different hydrocolloids (Figure 1). The close correlation of these models could have two primary origins:

- *either* the sensory perception of viscosity, corresponding to the oral shear stress, has a direct impact on flavor due to a perceptual interaction, *or*
- the Kokini oral shear stress predicts a physical effect (e.g. upon mass transport rates for sucrose), which is in turn responsible for the sweetness reduction.

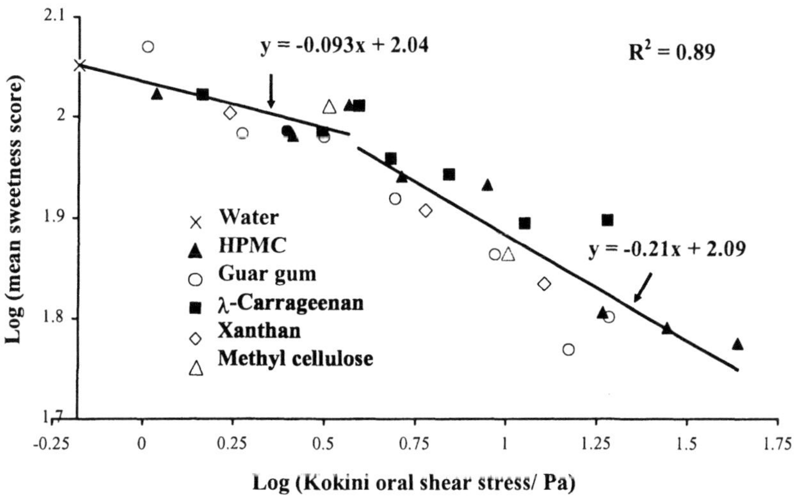

Figure 1. Model relating sweetness perception to the oral shear stress developed in mouth whilst drinking hydrocolloid solutions (all contained 50 g/L sucrose and 100 ppm isoamyl acetate).

In this paper, two experiments are described, which continue our investigation into factors influencing flavor perception in viscous solutions.

Aroma release from viscous solutions

Previous investigations (*3,10*) had concluded that aroma delivery in-nose was not substantially altered by increasing concentrations of hydrocolloid thickener. Each of these conclusions was based on the analysis of single aroma systems. To determine the generality of these findings, a more detailed investigation of aroma release from viscous hydrocolloid solutions was undertaken. Since all prior indications were that viscosity did not have considerable effects on aroma release, we adopted a top and bottom approach, where release from thickened solutions at 3.5 times the c* concentration (the most viscous solutions used in prior experiments) was compared to that from water. A test set of seven aroma compounds were selected to include a range of physicochemical properties, in particular with respect to volatility and hydrophobicity. Results presented here for λ-carrageenan were representative of those observed in parallel experiments with HPMC and guar gum.

Experimental

Sample preparation:
λ-carrageenan (17 g/L; Red Carnation Gums Ltd, Laindon, UK) was dispersed in water (60°C) using an overhead paddle stirrer at between 200-600 rpm. Once dispersed, the solution was cooled to 5°C and stirred for a further 6 hours to ensure adequate hydration of polymer chains. Volatiles were analyzed in two separate groups for both aqueous and thickened samples to minimize interference or suppression of ionization during API-MS analysis. Volatile concentrates (dispersed in propylene glycol) were diluted into water and λ-carrageenan samples to produce final concentrations which resulted in a similar ion intensity for each analyte (Table 1). The solutions were mixed on a roller bed for two hours to disperse volatiles adequately in the viscous matrix.

Table 1. Experimental detail. Sample preparation and analysis

Compound	Selected ion for API-MS (m/z)	Cone voltage (V)	Concentration in the samples (ppm)	
			Headspace experiments	Nosespace release
menthone	155	18	0.22	55.3
IAA	131	23	0.17	43.3
IBMP	167	25	0.35	87.2
ethanol	47	18	8	1990
oct-1-en-3-ol	111	10	4.15	51.8
diacetyl	87	20	7.84	71
benzaldehyde	107	22	0.62	15.6

IAA = isoamyl acetate; IBMP = 2-isobutyl-3-methoxy-pyrazine.

Headspace analysis:
Five 100g aliquots of each solution were weighed into separate 250ml Schott bottles, sealed with stoppered lids and allowed to equilibrate for 30 min. Samples were analyzed in random sequence, by removing the stopper from the lid and sampling headspace directly into the MS-Nose™ (Micromass, Manchester), using a sampling rate of 5mL/min. Operation was in selected ion mode, using the analytical conditions specified in Table 1. Volatile concentrations in the headspace were calculated relative to the signal for a calibrant of known concentration (*11*). Air-water partition coefficients were calculated for aroma compounds in each matrix, assuming that the liquid phase concentration remained as prepared (i.e. that loss of aroma to the headspace was negligible by comparison).

Nosespace release studies:
Aroma release was measured in-nose during consumption of the samples using the MS-Nose™ (sampling rate 30 mL/min). Two panelists consumed eight replicates of each sample, according to a fixed protocol. They were asked to breathe in, sip 8ml of solution from a spoon, close their mouth and swallow the sample, then exhale and continue to breathe normally whilst resting their nose on the MS-Nose™ nasal sampling tube. The maximum in-nose volatile concentration (I_{max}) was calculated for each drinking event by comparison with the signal for a calibrant of known concentration.

Results and Discussion

Headspace studies showed only minimal effects of 17 g/L λ-carrageenan on volatile partition behavior, relative to aqueous systems (Figure 2).

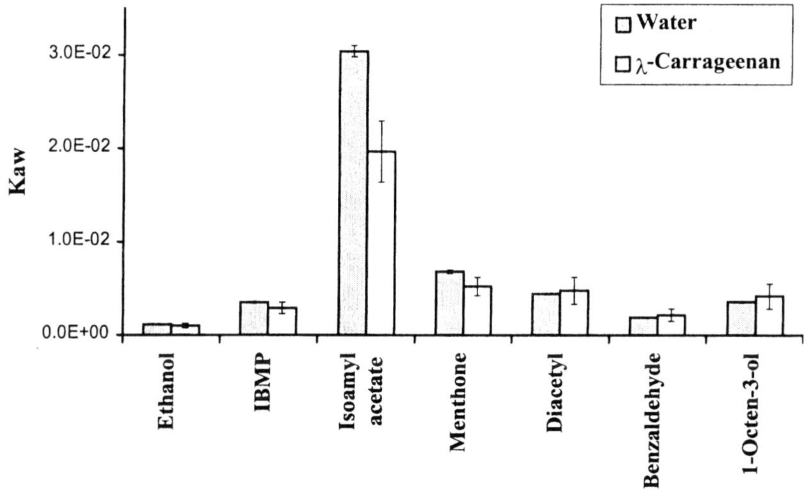

Figure 2: Air-water partition coefficients (24 °C) of seven volatiles in water and 17 g/l λ-carrageenan (mean of five replicate measurements)

The only significant differences in partition due to thickener were observed for isoamyl acetate and menthone, each of which had lower partition coefficients in λ-carrageenan (salting-in effect). Menthone was the most hydrophobic of the volatiles studied and it is possible that the mechanism for its retention in λ-carrageenan involves inclusion in hydrophobic folded regions of the polymer chain (*12*). Isoamyl acetate is relatively polar and the magnitude of the observed effect (30% reduction in headspace concentration) suggests a more specific interaction might occur between isoamyl acetate and λ-carrageenan.

Nosespace release studies (Figure 3) again showed relatively minor differences in aroma release between the two matrices, bearing in mind that we were comparing water with a carrageenan solution that had the consistency of a thick sauce (zero-shear viscosity ≈ 4 Pas).

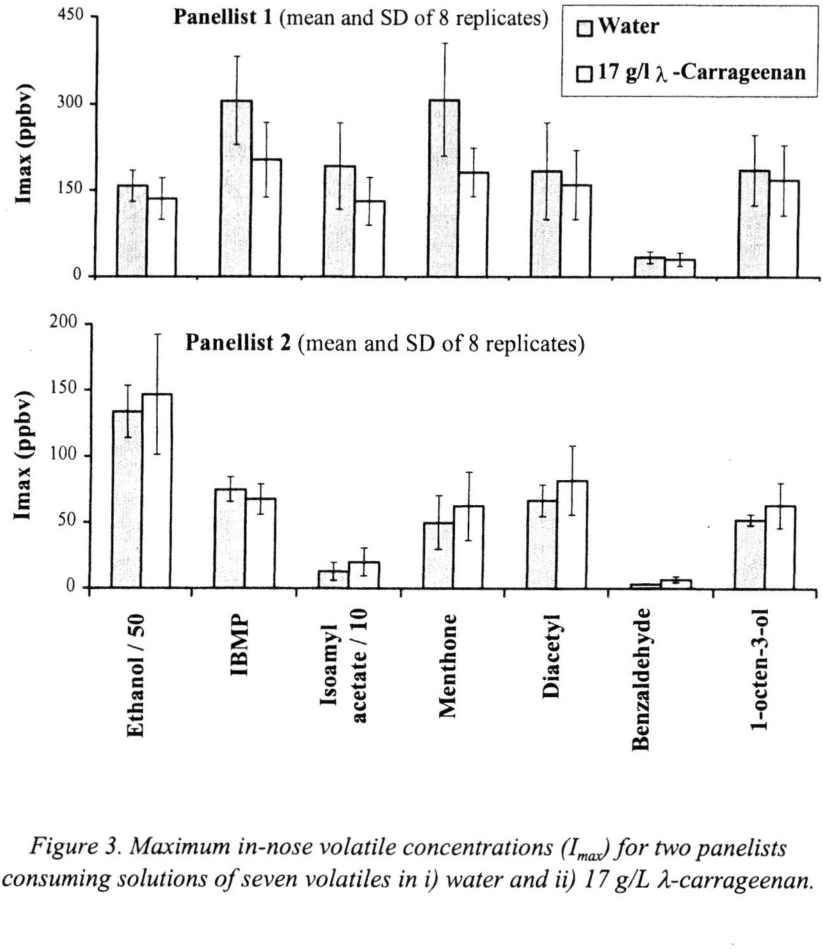

Figure 3. Maximum in-nose volatile concentrations (I_{max}) for two panelists consuming solutions of seven volatiles in i) water and ii) 17 g/L λ-carrageenan.

The principal source of variation was between panelists, with panelist 1 having I_{max} values up to four times as great as those observed for panelist 2. There were also different trends in aroma release for the two panelists. Panelist 1 on average showed higher aroma release from aqueous solutions, whilst panelist 2 tended to have higher release from thickened samples. This difference is probably attributable to the way in which panelists swallowed the samples. Factors such as swallowing pressure and the time taken for the viscous sample to clear the airways on the way to the esophagus would influence aroma delivery from viscous samples. Apparently such factors were more significant in determining aroma release than any effects due to viscosity itself.

This work confirmed the hypothesis that an increase in viscosity does not in itself cause a decrease in aroma release in-nose. It is interesting to note that observed decreases in static equilibrium headspace concentrations of menthone and isoamyl acetate did not result in significant reductions in nose-space release. This observation is consistent with reversible binding between the volatile and hydrocolloid, so that static, but not dynamic release was affected.

Perception of sweetness in model sauces thickened with starch and HPMC

This experiment investigated whether models based upon the Kokini oral shear stress could be used to predict sweetness perception in starch thickened solutions as well as for random coil-polysaccharide thickeners such as HPMC.

Experimental

Sample preparation:
HPMC samples were prepared as described previously (*1*). Starch 'sauces' were prepared by mixing starch powder (cornflour or arrowroot, purchased from J. Sainsburys supermarket, London, UK) to a thin paste with water. This paste was then dribbled on to the vortex of a beaker of water (85-90°C) being stirred at 300 rpm with an overhead paddle stirrer. Starch was added over a period of one minute and the solution maintained at 85-90°C for a further 3 minutes, with stirring at 400 rpm. The solution was then standardized gravimetrically with additional water, mixed, transferred to a Shott bottle and rolled on a roller bed (SRT2; Stuart Scientific, Redhill, UK) at 5°C to cool. Sucrose (80 g/L) was added once the starch was cool. The solutions were rolled continuously prior to

sensory evaluation to avoid lump formation. Pineapple aroma (100ppm ethyl 3-(methylthio)-propionate) was added and the final 'dessert sauce' rolled for 2 hours to ensure adequate mixing prior to sensory evaluation. The pineapple aroma was selected because screening headspace studies using the MS-Nose™ showed no obvious interactions with HPMC or either type of starch.

Sensory Evaluation:
A trained panel of 11 assessors (3 men & 8 women) rated each of 14 thickened samples twice for sweetness intensity over two panel sessions. Magnitude estimation against a fixed modulus was used to score each solution relative to the reference (sweetness = 100), which comprised 80g/L sucrose and 100ppm pineapple aroma in water. The reference was scored twice as a sample during each session.

Sample rheology:
The flow characteristics of each solution were measured at 25°C on a Bohlin CS-10 applied stress rheometer (Bohlin Instruments, Lund, Sweden). A linear fit to the shear-thinning portion of each flow curve was used to calculate the power law parameters required for the calculation of the Kokini oral shear stress presented by each sample (*10*). Since starch sauces showed time-dependent rheology, viscosity measurements were carried out in duplicate both before and after each panel session. The value used for the Kokini oral shear stress is the mean of these measurements.

Results and Discussion

A double logarithmic plot of mean sweetness perception against oral shear stress of the samples showed good correlation (Figure 4). Data for HPMC, cornflour and arrowroot could be fitted (R^2=0.88) with a Boolean function comprising two linear series in the low and high oral shear stress ranges (*10*). The transition between the two series occurred somewhere between an oral shear stress of 5-7 Pa. With the previous model (Figure 1) this transition occurred at around 4 Pa, although this was at a sucrose concentration of 50g/L. It seems logical that a greater oral shear might be required to start to inhibit sweetness at higher stimulus concentrations. Alternatively, the slight sweet taste of cornflour itself in solution might have influenced sweetness scores of some of the samples which were important in defining the break between the two series. This might also explain why the cornflour samples were generally rated a little sweeter than the model would have predicted.

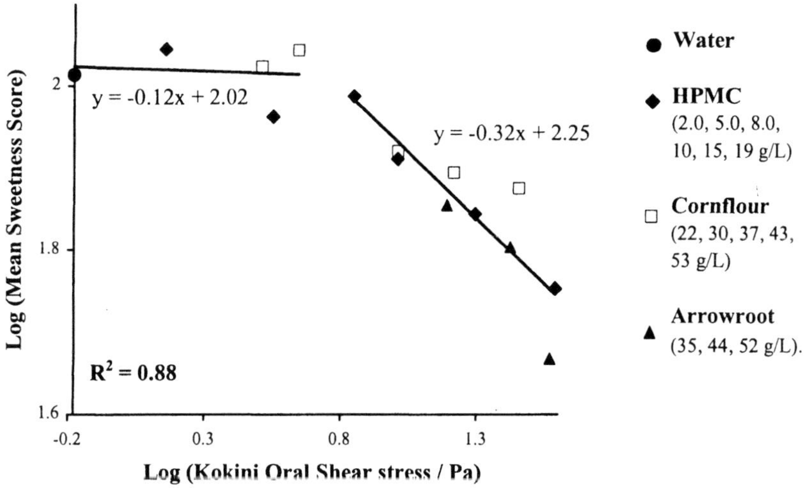

Figure 4: Sweetness perception as a function of the Kokini oral shear stress for pineapple 'dessert sauces'

It is noteworthy that a model based on the oral shear stress can predict the sweetness suppressing behaviors of starches and HPMC together, since the structure at a molecular level of these two systems will be quite distinct. This provides further support for the hypothesis that a sensory input for viscosity, proportionate to the oral shear stress, is interacting with sweetness perception.

Flavor as a multisensory percept

Everyday experience involves the continuous integration of information from multiple sensory inputs (*13*). There are clear advantages to be gained from combining sensory inputs to provide information about the environment which is not available from within a single modality. Flavor perception is no different and comprises the sensory combination and integration of odors, tastes, oral irritations, thermal sensations and mouthfeels that arise from a particular food (*14*).

The inter-dependency of the senses has been well documented in some areas of research (e.g. interactions between sound and vision, such as the ventriloquism and McGurk effects (*15*)). Within flavor research, there have been studies that demonstrated effects of vision on flavor perception (*16*) and a whole field of literature related to taste-aroma interactions. Dalton *et al.* (*17*) have recently shown co-integration of sub-threshold taste and aroma stimuli (detected when presented together at approximately 63% of their individual detection thresholds). This suggests direct neural integration of the two modalities.

The neural mechanism for cross-modal integration involves convergence of sensory streams at 'multimodal' neurons, which are responsive to more than one sensory modality. Electrophysiological studies have identified the presence of such neurons in, for example, the primate orbitofrontal cortex (*18*).

Commonality in gustatory and somatosensory perception.

Neurally, there is considerable overlap between the gustatory and somatosensory systems. The taste buds of the oral cavity are innervated by branches of three cranial nerves (VII, IX, X). Each of these nerves also carries somatosensory afferents that innervate regions of the tongue surrounding the taste buds. The presence of these somatosensory afferents makes it difficult to distinguish pure taste sensations from somatosensory information. The chorda tympani, for example, may respond to both taste and temperature or tactile stimulation (*19*).

Recent advances in magnetic resonance imaging have facilitated studies of the parts of the human brain that are involved in processing selected stimuli. Cerf-Ducastel *et al. (19)* investigated the interaction between gustatory and lingual somatosensory perception in the cortex. Hydrochloric acid (25 & 31 mM) and alum were used as pungent and astringent somatosensory stimuli respectively. It was concluded that taste and lingual somatosensory modalities appear thoroughly intermingled, both at a peripheral and a cortical level. Only subtle differences in brain activation were observed between pure taste and somato-gustatory (mixed) stimuli, indicating the complex task of analysis performed by the brain.

Why might viscosity affect sweetness perception universally ?

From neurological evidence it seems entirely feasible that viscosity interacts with and is itself a component of flavor perception. The common neural pathways and similar patterns of brain activation for gustatory and somatosensory stimuli together with studies identifying bimodal neurons that are responsive to taste and somatosensory stimuli (*20*) all support this hypothesis. This still raises the question: why might viscosity have a more universal effect on sweetness perception than on the other basic tastes, as seems to be the case from reviewing the literature (*1*).

Looking globally at the influence of somatosensory stimuli on taste, there is a pervasive influence upon sweetness perception. Cruz and Green (*21*) have shown that warming of the anterior tongue can induce sweetness, whilst cooling may stimulate sourness or saltiness. This is consistent with electrophysiological evidence showing thermal sensitivity in peripheral taste neurons (*22*). Temperature also interacts with sweet taste. Green and Frankmann (*23*) reported that the perceived sweetness of solutions of glucose, fructose and aspartame was significantly reduced by cooling the tongue. Interestingly, no such effect was observed with saccharin, possibly due to differences in taste transduction (*24*). The oral irritant capsaicin has been shown to influence taste perception when the stimuli are presented sequentially (*14*). Simons *et al.* (*25*) showed that pre-treatment of one half of the tongue with capsaicin resulted in reduced perception of sweetness (sucrose), quinine induced bitterness and umami taste. In reviewing the literature regarding effects of capsaicin on taste perception, the authors commented that there is a general consensus that capsaicin reduces perceived sweetness, whilst its effect on other modalities has differed, dependent upon methodological issues. It was concluded that the effects of capsaicin on taste might be caused both by peripheral effects upon G-protein coupled taste transduction mechanisms and by central neural interactions.

Cowart (*26*) investigated the effects of carbonation on perception of the basic tastes. Carbonation resulted in decreased sweetness and saltiness and increased sourness. As discussed by the author, these results may reflect trigeminal-gustatory interactions, but could also be explained by direct taste effects of CO_2 in solution.

Conclusions

Somatosensory stimuli interact with gustatory stimuli at both peripheral and central neural levels. Researchers have shown that thermal and irritational stimuli interact with sweet taste. Viscosity, as a tactile stimulus, appears to act likewise. The generality of the effect of viscosity on perceived sweetness, but not on other taste modalities, may result from specific neural interactions and features common to the transduction of all sweet agents. The fact that it is easier to find agents which will suppress the sweetness of a whole range of sweet molecules (e.g. lactisole, *Gymnema Silvestre*) than is the case for bitterness (*14*), suggests this commonality.

Whilst somatosensory-taste interactions may account for general reductions in perceived sweetness observed above an oral shear stress of 4-5 Pa, other effects of viscosifying agents may still influence the sweetness of specific systems. These include taste qualities of the thickener itself, reductions in the diffusion rates of taste molecules through viscous matrices (*27*) and physical barrier effects which might hinder access to taste receptors at high viscosities. Experiments conducted in our laboratories (*3,10*), including the current study, have all shown small reductions in sweetness in the dilute thickener range (e.g. Figures 1 & 4). This may be the 'background' rate of inhibition due to some of these other factors, before a multi-modal neural mechanism commences when the oral shear stress stimulus is sufficiently high. A combination of factors in this manner might explain why the influence of thickeners on sweetness perception has proved a complex area of study.

It is unlikely that the influence of viscosity on diffusion rates is a key determinant of the taste intensity of viscous solutions, since this theory fails to account for observed differences between the modalities; e.g. bitter taste is not universally affected by viscosity, even though diffusion effects would be predicted to be equally prevalent on bitter and sweet molecules of a similar size. Furthermore, results presented here have shown that the mass transport of aroma molecules to the olfactory receptors is unaffected at viscosities considerably above those at which flavor suppression occurs. It can be concluded that, aside from specific situations where there are physico-chemical interactions between thickener and aroma molecule, changes in flavor perception with increasing viscosity are driven by changes in taste perception.

When modifying the texture of viscous food systems, it is important from a flavor standpoint to consider these changes in non-volatile taste, since the non-volatiles create the psychological-sensory foundation of flavor on which the volatiles build (*14*).

References

1. Cook, D. J.; Hollowood, T. A.; Linforth, R. S. T.; Taylor, A. J. *Food. Qual. Prefer.* **2002**, *in press.*
2. Baines, Z. V.; Morris, E. R. *Food Hydrocolloids* **1987**, *3*, 197-205.
3. Hollowood, T. A.; Linforth, R. S. T.; Taylor, A. J. *Chem. Senses* **2002**, *27*, 583-591.
4. Noble, A. C. *Trends Food Sci. Technol.* **1996**, *7*, 439-444.
5. Christensen, C. M.; Casper, L. M. *Journal of Food Science* 1987, *52*, 445-447.
6. Shama, F.; Sherman, P. *J. Texture Stud.* **1973**, *4*, 111-118.
7. Wood, F. W. In *Rheology and texture of foodstuffs*, **1968**; Vol. S.C.I. Monograph No.27, p 40.
8. Kokini, J. L. *Food Technology* **1985**, *39*, 86-&.
9. Elejalde, C. C.; Kokini, J. L. *Journal of Texture Studies* **1992**, *23*, 315-336.
10. Cook, D. J.; Hollowood, T. A.; Linforth, R. S. T.; Taylor, A. J. *Chem. Senses* **2002**, *In press.*

11. Taylor, A. J.; Linforth, R. S. T.; Harvey, B. A.; Blake, A. *Food Chemistry* **2000**, *71*, 327-338.
12. Roberts, D. D.; Elmore, J. S.; Langley, K. R.; Bakker, J. *Journal of Agricultural and Food Chemistry* **1996**, *44*, 1321-1326.
13. Calvert, G. A.; Brammer, M. J.; Iversen, S. D. *Trends in Cognitive Sciences* **1998**, *2*, 247-253.
14. Breslin, P. A. S. *Flavour and Fragrance Journal* **2001**, *16*, 439-456.
15. McGurk, H.; MacDonald, J. *Nature* **1976**, *264*, 746-748.
16. DuBose, C. N.; Cardello, A. V.; Maller, O. *J.Food.Sci.* **1980**, *45*, 1393-1399.
17. Dalton, P.; Doolittle, N.; Nagata, H.; Breslin, P. A. S. *Nature Neuroscience* **2000**, *3*, 431-432.
18. Rolls, E. T.; Baylis, L. L. *Journal of Neuroscience* **1994**, *14*, 5437-5452.
19. Cerf-Ducastel, B.; Ven de Moortele, P. F.; MacLeod, P.; Le Bihan, D.; Faurion, A. *Chemical Senses* **2001**, *26*, 371-383.
20. Travers, S. P.; Pfaffmann, C.; Norgren, R. *Brain Res.* 1986, *365*, 305-320.
21. Cruz, A.; Green, B. G. *Nature* **2000**, *403*, 889-892.
22. Ogawa, H.; Sato, M.; Yamashita, S. *J. Physiol. (Lond.)* **1968**, *199*, 223-240.
23. Green, B. G.; Frankmann, S. P. *Physiology & Behavior* **1988**, *43*, 515-519.
24. Brand, J. G.; Feigin, A. M. *Food Chemistry* **1996**, *56*, 199-207.
25. Simons, C. T.; O'Mahony, M.; Carstens, E. *Chemical Senses* **2002**, *27*, 353-365.
26. Cowart, B. J. *Chemical Senses* **1998**, *23*, 397-402.
27. Cussler, E. L.; Kokini, J. L.; Weinheimer, R. L.; Moskowitz, H. R. *Food Technology* **1979**, *33*, 89-92.

Chapter 17

Effects of High-Fructose Corn Syrup on Perception and Release of Flavors in Soft Drinks

Thomas N. Asquith[1,3], and Robert L. Swaine, Jr.[2]

[1]Asquith Flavor Systems, Cincinnati, OH, 45231
[2]Proter & Gamble Company, 6210 Center Hill Avenue, Cincinnati, OH 45224
[3]Current Address: Brown-Forman Distillary Company, 850 Dixie Highway, Louisville, KY 40210

Small changes in the amount of carbohydrate sweetener (Brix) can totally change the sensory character of a soft drink. Are the sensory effects caused by changes in release and/or perception of flavor? Model soft drinks were prepared that differed by 1 g of high fructose corn syrup per 100 g of beverage. The acid/brix ratio, amount of flavor and other ingredients were constant. The composition of head-space and breath were analyzed by APCI-MS. Replicate samples were rated by sensory panels for both aroma and flavor. The sensory ratings for changes in overall flavor and sweetness were larger than for changes in aroma. The composition of headspace and breath were changed by Brix. We concluded that Brix had more impact on flavor perception than on flavor release.

The amount of sugar (Brix) in juices and soft drinks has a major impact on consumer ratings for flavor. A difference of only one Brix (about 1% w/w) can significantly change the flavor of beverages. The cause for this flavor change is not fully understood. Possible explanations are changes in flavor perception and/or flavor release. Brix does affect the perception (*1,2*) and the release of flavor volatiles (3, 4, 5).

However these studies reported the effects of large changes in Brix on flavor. Companics typically make small changes in Brix when developing new formulas. Yet flavorists must still adjust flavor composition to compensate for changes in Brix. Often this work relies on repetitive cycles of formulation and sensory evaluations. Understanding the impact of formulation changes on flavor systems should reduce development time and expense. Our experiments assessed the impact of small changes in Brix on perception and release of flavor. The experiments were done as part of a product development project to launch new varieties of juice products on accelerated timelines.

Materials and Methods

Sensory Evaluations

Panelists were prescreened for ability to recognize and distinguish between basic tastes and aromas. They were also tested for ability to rank order the intensities of tastes and aromas on a nine point scale. Panelists also regularly consumed juice products. Samples were rated by thirty to fifty panelists for overall flavor intensity, sweetness, tartness, uncharacteristic flavor, aftertaste and overall liking. The rating scales ranged from 1 = low intensity to 9 = high intensity. Panelists were instructed to evaluate first for aroma and then for flavor.

Instrument and Operating Conditions

Spectra were collected using a Micromass ZMD under conditions for atmospheric pressure chemical ionization (APCI). For headspace analyses the capillary flow rate was 2 ml/min with a desolvation gas flow rate of about 600 l/h. Capillary voltage was 4.50 kV, extractor voltage was 3.0 V, Rf lens was 0.3, source block temperature and desolvation gas temperatures were 50°C. The transfer line was a fused silica capillary (0.53 mm ID) encased in a copper line, which was kept at 120°C. Individual flavor components were analyzed by total ion current to verify major ions and relative purity. Ionization efficiency of compounds depended on cone voltage values. Therefore each compound was analyzed as single ions (SIM-mode) to determine optimal cone voltages for maximum ionization efficiency (Table 1). Cone voltage values were generally higher for breath analysis than for headspace analysis. Note that mass resolution of the Micromass ZMD was too small to distinguish between the common terpenes in orange juice. Although method development was done with limonene and linalool the experimental values were reported as generic "terpenes". For actual analyses of samples, identities were based on abundance and behavior relative to standards. Identities were not confirmed by spiking standards. Conditions for breath analyses were based on recommendations from Dr. R. Linforth (personal communication). The capillary flow rate was about 280 ml/min with a desolvation gas flow rate of about 710 l/h. The high flow required the 1 mm (ID) tip rather than the 1.6 mm (ID) tip. Other parameters remained the same. Flavor components were analyzed as single ions (SIM-mode).

Headspace Analysis

The basic method was adapted from (*3*). Fifty ml of sample was poured into a 100 ml Schott Bottle and equilibrated at room temperature for at least two hours. Caps for the Schott bottles had holes of about 3 mm. At ten minute intervals, the capillary was inserted about two centimeters into the jar. Headspace was sampled for one minute. Each sample was analyzed with seven replications. Excel software was used to calculate values for average peak area, standard deviation and coefficient of variation.

Breath Analysis

Fifty mL of beverage were poured into a plastic cup. The breath was sampled by exhalation through a plastic tube into a T-splitter (Swagelock), which was connected to the transferline. Representative baselines were established by monitoring the acetone (M H+ = 59) peak in breath during several exhalations. Then beverages were consumed by sipping through a straw. After each swallow, the released odorants were exhaled. No further standardization was performed. Each beverage was sampled twice. There were large variations in the absolute amounts of ion from breath to breath. However the relative amounts of ions are fairly constant breath to breath (*6*). Seven distinct breaths were chosen. Using values for ion 117 as the common denominator, the other ions were converted to ratios against ion 117.

Table I: MH+ Values and Cone Voltages for Selected Compounds

Compound	*Major Ion*	*C. V. HS*	*C.V. NS*
Acetone	59	NA	25
Ethyl Butyrate	117	18	18
Ethyl Valerate	131	16	18
Benzaldehyde	107	24	26
Octanal	129	19	21
Methyl Anthranilate	151	14	16
Limonene	155	20	22

Samples

The "fruit punch" base contained a blend of fruit flavors, gum arabic, colorants and sodium benzoate. The amount of high fructose corn syrup (HFCS) was adjusted such that samples contained 11, 12, 13 or 14 Brix HFCS. The amount of citric acid was adjusted such that there was a constant acid/Brix ratio of 0.011. The citrus base contained citrus oils supplemented with individual flavor compounds as well as sodium hexametaphosphate, colorants and potassium sorbate. The amount of high fructose corn syrup (HFCS) was adjusted such that samples contained 11, 12 or 13 Brix HFCS. The amount of citric acid was adjusted with Brix such that there was a constant acid/Brix ratio of 0.03.

Results

Sensory

Ratings for aroma intensity were essentially constant (Figure 1). In contrast ratings for flavor intensity, sweetness and hedonics increased as Brix increased. Sourness intensity tended to parallel sweetness but the percent change was less than for sweetness (data not shown).

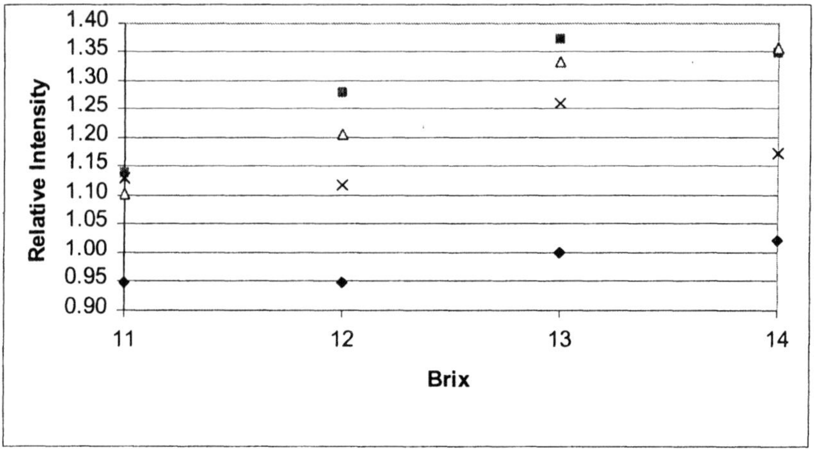

Figure 1. Effect of Brix on Sensory Attribute; ♦ = aroma intensity, ■ = flavor intensity, ▲= sweetness intensity and X = overall liking

Volatile Analyses

Method Precision: coefficient of variation values were less than 0.05 for ions 117, 129 and 131, and generally less than 0.10 for ions 107, 151 and 155. Day to day variation was less than 10% except for ion 155, which varied by 20%.

Effect of HFCS on Headspace Composition: Values from the fruit punch product are shown in Figures 2 and 3. The amounts of ions 117, 129 and 131 changed by about 20% between 11 to 14 Brix (Figure 2). The amounts of ions 107, 151 and 155 changed by about 70% between 11 to 14 Brix (Figure 3). The standard deviation values were small enough to detect significant differences in headspace composition between products that differed by one Brix. Clearly there was a larger effect of Brix on ions 107, 151 and 155 than on the other ions. The amount of Brix did not change the amounts of ions 117, 129 and 155 in headspace of the citrus product (data not shown). In order to check if the composition varied due to different amounts of flavor in the products, average peak area values were converted to ratios using ethyl butyrate as the common denominator (Table II). The trends were similar to those in Figures 2 and 3.

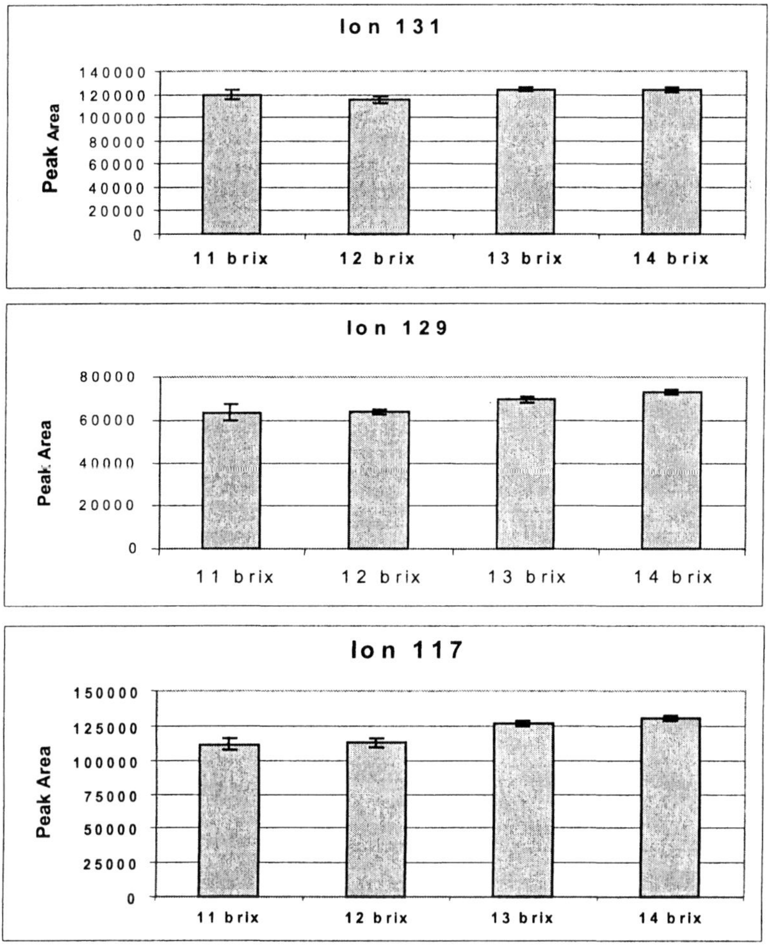

Figure 2: Effect of Brix on Ions 117, 129 and 131

Table II: Ratios of Volatiles in Headspace

Brix	*107/117*	*129/117*	*131/117*	*151/117*	*155/117*
11	11.9	0.6	1.1	1.5	14.6
12	12.4	0.5	1.0	1.6	13.4
13	13.8	0.5	1.0	2.0	18.6
14	16.0	0.6	1.0	2.3	21.0

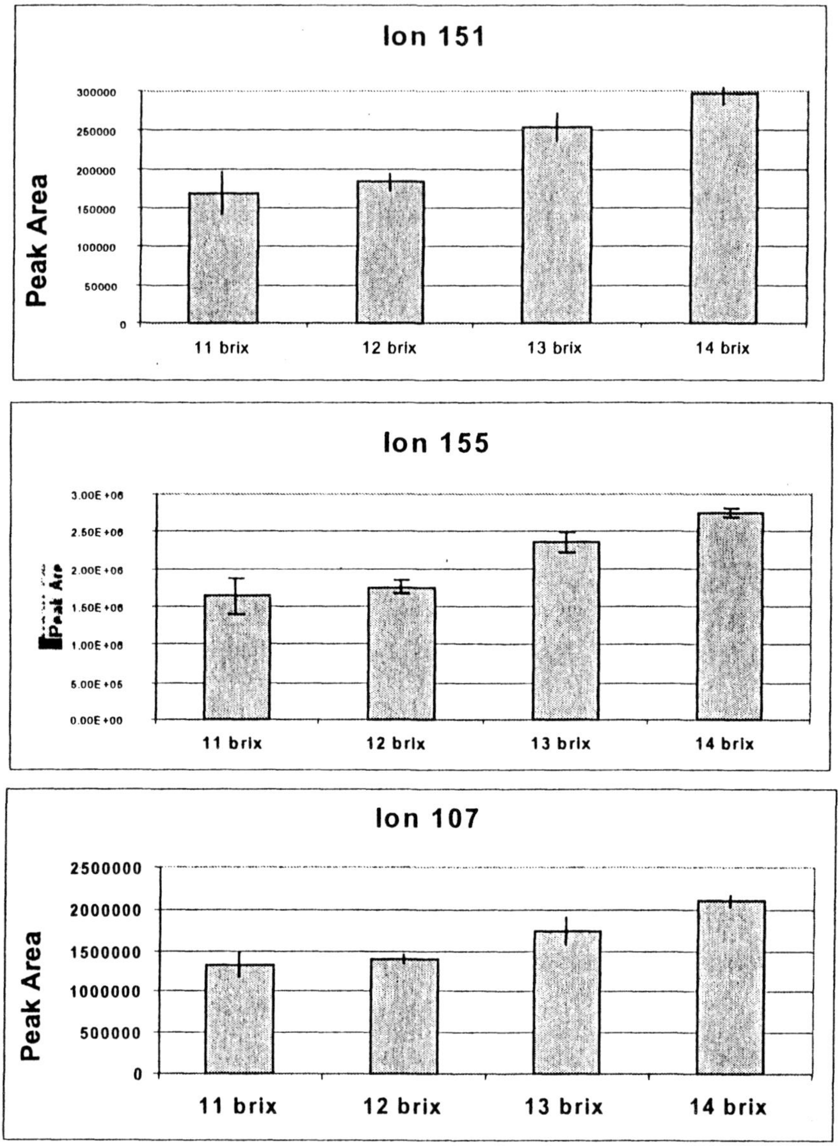

Figure 3: Effect of Brix on Ions 107, 151 and 155

Effect of HFCS on Breath Composition: Average peak area values were converted to ratios using ethyl butyrate as the common denominator (Table III). Values for 129/117 and 131/117 were essentially unchanged, 107/117 was variable while 151/117 and 155/117 increased as Brix increased. The trends were similar to those in Table 2. Just as with headspace composition, there were few differences between 11 and 12 Brix. Interestingly there was relatively more ion 151, and less ion 155, in the breath than in the headspace. The relative decrease in ion 155 was previously observed when the compositions of headspace and breath were analyzed for citrus products (*6*).

Table III: Ratios of volatiles in breath.

Brix	*107/117*	*129/117*	*131/117*	*151/117*	*155/117*
11	13.6	0.3	1.6	3.5	3.4
12	12.6	0.4	1.1	3.0	3.4
13	11.4	0.4	1.7	4.0	4.0
14	15.5	0.4	1.4	5.4	4.9

Conclusions

APCI-MS can provide accurate measurements of headspace composition. The measurements showed that a difference of one Brix can affect headspace composition. Terpenes and aromatic compounds were more affected than esters and short chain aldehydes. Our results contrasted with published work (*3, 4, 5*). This was possibly because we examined fairly low concentrations of sugars (about 10-15%) whereas others examined effects across a broader range (about 20-60%). Changes in headspace composition reflect changes in water structure and solute solubility (*3, 4, 5*). Solute solubility may be very different at lower concentrations of sugar. Flavor solubility in ethanol/water systems is complex because the physical chemical structure of ethanol/water mixtures changes as the relative amounts of the two solvents change (*7*). Flavor release can be enhanced or suppressed depending upon percent alcohol and nature of the solute.

Interestingly the same effects on flavor release were generally observed in exhaled breath after drinking products. Therefore the basic principles of release were similar in both the bottle and in the mouth. The relative decrease of terpenes in the breath, versus in the headspace, is probably not related to Brix. We consistently detected this phenomenon in citrus products regardless of amount or type of sweetener (*6*). Loss of terpenes may result from dispersion of limonene into the saliva rather than less release into the vapor phase.

Despite the changes in headspace composition, panelists had a stronger reaction to the sweetness than to changes in aroma. The dominance of taste over aroma when rating beverage attributes was consistent with other work (*1, 2*). This suggests that adjusting total sweetness can compensate for small changes in Brix. Large changes in Brix may require targeted reformulation of flavor volatiles.

References

1. Nahon, D.; Koren, P.; Roozen, J.; Posthumus, M. *J. Agric. Food Chem.* **1998**, *46*, 4963-4968.
2. Davidson J.; Linforth R.; Hollowood T.; Taylor A. *J. Agric. Food Chem.*, **1999**, *47*, 4336-4340.
3. Friel, E.; Linforth, R.; Taylor, A. *Food Chem.* **2000**, *71*, 309-317.
4. Voilley, A.; Simatos, D.; Loncin, M. Lebensmitt. Wiss. Technol., **1977**, *10*, 45-49.
5. Roberts, D.; Elmore, J.; Langley, K.; Bakker, J. *J. Agric. Food Chem.*, **1996**, *44*, 1321-1326.
6. Zehentbauer, G.; Asquith, T.; Li, J.-J. in *Trends in Flavour Research*; Editor Etievant, P., Elsevier Science, Amsterdam, **2003**; in press.
7. Conner, J.; Paterson, A.; Piggott, J. *J. Sci. Food Agric.*, **1994**, *66*, 45-53.

Chapter 18

The Role of Esterified Compounds in the Development of Staleness in Fresh-Cut Fruit

Olusola Lamikanra

Southern Regional Research Center, Agricultural Research Service, U.S. Department of Agriculture, 1100 Robert E. Lee Boulevard, New Orleans, LA 70124

The aroma and taste of most fruits are considerably influenced by esters. Fresh-cut fruit products have considerably shorter shelf life than the intact fruit. Changes in volatile esters of fresh-cut cantaloupe melon and pineapple were determined by headspace solid-phase microextraction gas chromatography – mass spectrometry (SPME GC-MS). Several volatile ester compounds including methylbutyl acetate, hexyl acetate, ethylisobutyrate, methyl 2-methylbutyrate, ethyl butyrate and ethyl 2-methylbutyrate were present in cantaloupe melon while methyl 2-methylbutanoate, methyl hexanoate, methyl 5-hexenoate, ethyl hexanoate and ethyl 5-hexenoate were the dominant esters identified in pineapple. Storage at 4 °C for a period of 24 h. caused significant decreases in total volatiles, aliphatic and aromatic esters in both fruits. Cantaloupe melon hybrids with extended uncut postharvest shelf life had lower amounts of esters than the *Mission* cultivar. The lack of correlation between postharvest shelf life of whole cantaloupe and the fresh-cut fruit is indicated by the similar loss in esters during storage of cut fruit from the hybrids. Esterase activity measurements in cantaloupe suggest that the catalytic effect of the enzyme occurs very early after fresh-cut processing.

Introduction

Fruit flavor is the result of a delicate balance of volatile and non-volatile compounds. The volatile compounds detected by the olfactory nerve endings in the nose are major determinants of fruit quality as perceived by consumers. In cut fruits, physiological changes as a consequence of tissue exposure that affect flavor include browning, accelerated water loss and increased water activity. Enzymatic activities increase as a result of the increased permeability that results from tissue disruption and mixing of enzymes and substrates that are otherwise sequestered within vacuoles. Carbon supply from freed soluble sugars also enhances potential microbial attack.

Cut fruit products rapidly loose their typical flavor, even when stored under refrigerated conditions. It is well known that cut fruit could develop staleness or loss of freshness within a day of refrigerated storage. Esters, particularly those with acetate as the acyl portion, constitute one of the most important classes of compounds that impact the aroma of fruits. These compounds are generally believed to be primarily formed enzymatically by way of the combination of alcohols with acyl groups such as acetyl CoA during fruit development and ripening, under strict biosynthetic control by esterase (*1,2*). In this paper, the potential involvement of esterified compounds and esterase-mediated reactions in the loss of freshness of stored fresh-cut fruits is discussed.

Materials and Methods

Fruit preparation

Cantaloupe melon (*Cucumis melo* L. var. *reticulates*) and pineapple fruit were surface sterilized in bleach solution (10%) after which they were sliced longitudinally into two halves. Equatorial slices (~1- 2mm thick) were then cut on a Mandolin Kitchen Slicer (Robinson Knife Co. Buffalo, NY). After cutting off approximately 2mm along the slice edges, the fruit (3 g) was immediately chopped and the slurry was transferred into a vial (20 mL) containing NaCl (1 g) and into which a magnetic stirring bar was inserted. The vial was fitted with an aluminum septa cap and sealed. Benzothiophene, the internal standard used for GC-MS analysis dissolved in methanol was injected onto the pulverized fruit and mixed thoroughly by agitation. Fruit prepared for storage were sliced and placed in glass petri dishes at 4 and 22 °C respectively. Replicate samples from three separate fruits were removed and prepared for analysis at various times as described.

GC-MS Analysis

Volatile components of the fruit were extracted by headspace solid-phase microextraction (SPME) using a fused silica fiber coated with a 100μm layer of dimethylpolysiloxane. The fiber was conditioned by inserting it into the GC inlet for 2 h prior to use for volatile compound adsorption. The fruit and NaCl mixture was initially stirred in a water bath maintained at 30 °C for 30 min. The SPME fiber was then inserted into the sample headspace for 15 minutes while stirring continued at the same temperature. Desorption of the fiber took place in the GC inlet at 250 °C for 4 min.

GC-MS analysis was performed on a Hewlett Packard HP-6890 series system utilizing a HP-5 MS Crosslink 5% Phenyl Methyl Siloxane (30m x .25mm x .25μm) column. The injection port was operated in a splitless mode with helium as the carrier gas. The oven was programmed at an initial temperature of 60 °C ramped to 215 °C at the rate of 8 °C/min, then to 260 °C at °C/min and held for 15 min. The mass spectrometer was operated in a scan mode from 40 to 400 amu, using 70 eV electrons for ionization. Compounds were identified from their retention times using a commercially available library, authentic reference compounds, and MS fragmentation patterns. Relative amounts of the compounds recovered are expressed as peak areas relative to the peak area of the internal standard.

Determination of total proteins

Protein content of extracts was determined using bovine serum albumin standards and the Bio-Rad assay dye reagent. The diluted reagent in deionized water (20%; 5 mL) was added to 1 mL of standard solutions (0-1.5 mg/mL) and mixed thoroughly. Absorbance at 595 nm of the solutions was read after 15 min. Protein content of extracts was determined from a standard curve after they were reacted with the dye in a similar procedure.

Esterase Assay

On each sampling date, Tris buffer (pH 7.8, 0.05M, 80 mL) was added to the fruit (40g) and blended in a Waring blender for 1min. The mixture was then homogenized with a Tekmar Tissuemizer for 1 min. and centrifuged at 4 °C and 5,000 x g for 20 min. The supernatant and added ammonium sulfate (50%) was stirred at 4 °C for 30 min. after which it was placed in a freezer (-18 °C) for 1 h. The mixture was then centrifuged at 4 °C and 12,000 x g for 30 min. The supernatant was discarded and the residue homogenized in the Tris buffer (4 mL) for 1 min. This mixture was centrifuged at 4 °C and 12,000 x g for 30 min. The supernatant was used in assays for enzymatic activities. Esterase activity was monitored spectophometrically by measuring the production on *p*-nitrophenol from *p*-nitrophenyl acetate substrate. The reaction mixture (3 mL) contained deionized water (2.4 mL) *p*-nitrophenyl acetate (300 μL) and the enzyme extract (300 μL). *p*-nitrophenyl acetate (10 mM) was dissolved in dimethyl sulfoxide (DMSO). The reaction was started by addition of the enzyme extract. The solution was then incubated in a waterbath at 50 °C for 5 min. and centrifuged at 3,000 x g for 3 min. Absorbance was determined at 420

Results and Discussion

Most of the compounds identified in the fruit with the SPME method used are esters (Table I). Storage of cut cantaloupe fruit over a period of 24 h at 4 °C caused a sharp decrease in the total volatiles (Figure 1). The most drastic drop occurred in the relative amount of aliphatic esters in the fruit. After the initial decrease in the aliphatic and aromatic esters during the first 24 h, the loss of these volatile aroma compounds with storage time was minimal over a period of seven days (Table II). Concurrent with the decrease in the concentration of esters was the formation of terpenoid compounds; β-ionone and geranylacetone.

The reduction of esters appears to be an important early reaction step in the loss of freshness during storage of fresh-cut cantaloupe that could potentially serve as precursor compounds for synthesis of secondary volatile aroma compounds. A similar decrease in aliphatic esters, and synthesis of terpenoid compounds occur as a result of UV light induced stress in fresh-cut cantaloupe *(3)*. The concentration of aliphatic esters in fruit exposed to UV light for 15 min. decreased by over 60%. Cyclic and acyclic terpenoids, including phytoalexin compounds β-ionone, geranylacetone and terpinyl acetate were also produced. These compounds, particularly β-ionone were effective in inhibiting microbial growth in the fruit. It seems obvious from the similar pattern of changes in volatile compounds caused by UV light induced biological stress, as a consequence of storage, that the defense response of the fruit tissue plays a critical role in altering the flavor of the cut fruit. These changes in amounts of esters and terpenoid compounds are expected to affect fruit flavor. Esters are important flavor compounds in cantaloupe while terpenoid compounds such as β-ionone and geranylacetone have characteristic aroma properties *(4,5)*.

Esters are known to be involved in stress adaptation to tissue wounding *(6,7)* and those found in essential oils exhibit antifungal and antibacterial activities *(8)*. Tissue injury may cause esters that are otherwise inactive to be enzymatically converted into the aldehydes and alcohols with antimicrobial properties *(9)*. The compounds formed are responsible for the bitter and astringent tastes that are common when the fruiting bodies are broken. In many fruits, the natural defense mechanism against microbial infections provided by tannins also appear to be associated with the hydrolysis of ester linkages between gallic acid and polyols *(10)*. Japanese pear infected by *Venturia nashicola* is accompanied by de-esterification *(11)*. The stress-induced hydrolysis of polygalacturonide esters by pectin methylesterase that takes place during solubilization of the cell wall is well known *(12,13)*. The storage-induced reduction of ester concentration in cut cantaloupe is unrelated to microbial stress. Changes in volatile aroma compounds in cantaloupe treated with sodium azide were similar to the control that was not treated with the biocide *(14)*. The treated fruit, however, produced higher amounts of terpenoid compounds during storage.

Table I. Volatile compounds recovered in fresh-cut cantaloupe melon. Amounts are relative to the area of the internal standard. Peaks correspond to those identified in Figure 1. Reproduced from ref. 14.

Peak No	Compound	Relative Amounts*
1	Ethyl acctatc	1.95
2	Ethyl propionate	1.21
3	Ethyl isobutyrate	2.06
4	Ethyl - 2methyl butyrate	2.05
5	2 - Methyl butyl acetate	9.92
6	Methyl hexanoate	0.39
7	Ethyl (methylthio) acetate	0.32
8	Ethyl hexanoate	1.37
9	Hexyl acetate	3.6
10	1, 8-Cineole	0.08
11	1,3-Butanediol diacetate	0.34
12	2,3-Butanediol diacetate	0.13
13	Ethyl heptanoate	0.08
14	Benzyl acetate	1.73
15	Ethyl octanoate	0.09
16	Benzothiopene (standard)	1
17	Octyl acetate	0.27
18	Ethyl phenylacetate	0.06
19	Phenylethyl acetate	0.16
20	Phenyl hexanoate	0.06
21	3-Phenylpropyl acetate	0.16
22	Geranylacetone	0.01
23	β-Ionone	0
24	Dihydroactinidiolide	0.01
25	Diethyl phthalate	0.05

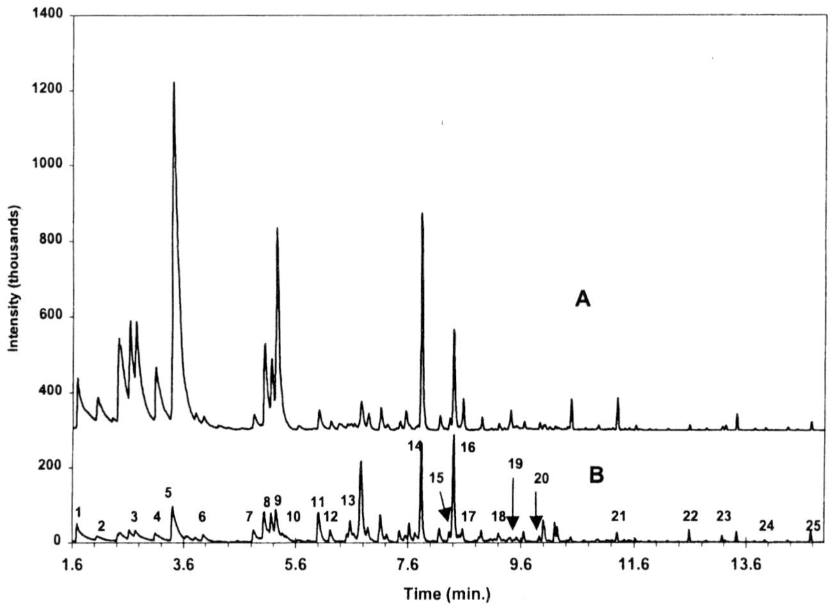

Figure I. . Total ion chromatogram of aroma compounds in fresh-cut cantaloupe melon (A) and the cut fruit stored at 4 °C for 24 h (B). Reproduced from ref. 14.

Plant tissue stress induced enzymatic hydrolysis of esters involves their esterase-mediated conversions to acids and alcohol *(9)*. Subsequent reactions involve fatty acid degrading enzymes and/or alcohol dehydrogenase. Most primary aroma compounds in cut fruits are believed to be products of β-oxidation of fatty acids and secondary compounds the result of fatty acid oxidation via the lipoxygenase pathway *(2)*. Many of the natural volatile compounds that control microbial growth in fruits are also typical products of the lipoxygenase reaction pathway *(10)*. Efforts aimed at correlating the reduction in aliphatic and aromatic esters during storage with synthesis of fatty acids, aldehydes and alcohols by solid phase microextraction of extracting the volatile compounds at higher temperatures (45 and 60 °C) were unsuccessful. While the extraction of volatiles at higher temperatures facilitated detection of many of the heavier compounds, including fatty acids, aldehydes and alcohols, the extraction procedure significantly decreased the amount of esters and increased terpenoid concentrations in the freshly processed fruit. The new compounds emitted under this condition could not be quantitatively correlated with the effect of storage on the ester compounds. This is apparently because of the ability of end products of hydrolysis of esters to serve as precursors in the formation of other fatty acid and their esters, alcohols and aldehydes, and such reactions increase with increased extraction temperature *(15,16)*.

Most traditional cantaloupe melon cultivars shipped across their regional planting locations (western shipper) have in their parentage *Mission* and *Haymark* or other similar cultivars with similar characteristics *(17)*. These hybrids are considered to be Traditional Shelf Life (TSL) fruits. The slip stage determines commercial maturity (usually ½ to ¾ slip) and, depending on the cultivar, the external color at this stage may retain a greenish cast. Skin color, however, typically change from gray to dull green when immature, deep uniform green at maturity, and light yellow at full ripeness. Although the postharvest shelf life of the cantaloupes varies according to harvesting stage (½ to ¾ slip) and handling, it is generally in the range of 12-16 days *(18)*.

Table II. Effect of storage at 4 °C on relative amounts of aliphatic and aromatic esters, and terpenoids present in fresh cut cantaloupe melon. Values are reported as amounts relative to that of the internal standard used.

Compounds	Fresh-cut	Days in Storage		
		1	3	7
Aliphatic Esters	23.43a	4.21b	2.6c	1.79c
Aromatic Esters	2.11a	1.04b	0.88bc	0.47c
Terpenoids	0.01a	0.1b	0.06bc	0.05c

Recent breeding efforts have produced hybrids that are more resistant to disease and insect damage *(19,20)*. In some of the cases these hybrids, at full ripeness, do not slip like the traditional cantaloupe melons and typically retain the green skin color *(7)*. This class of cantaloupe melon normally has a period of good postharvest storage life that can be as long as 20-25 days. The obvious characteristic that separates these non-slip (NS) fruits is the lack of abscission layer formed at the stem attachment when they reach maturity. Volatile aroma profiles of 6 NS hybrids were compared with that of the *Mission* cantaloupe melon *(21)*). The volatile and ester retention capability of the NS fruit relative to that of *Mission* was also determined. The results indicate that extended postharvest shelf life melons are more likely to have lower volatile aroma contents, and consequently reduced flavor complexity. The non-slip fruits and the *Mission* cultivar all contained methylbutyl acetate, benzyl acetate and hexyl acetate. Methylbutyl acetate and hexyl acetate are presumed to be the most important flavor compounds in *Cucumis melo* L, *(22)*. The lack of correlation between the postharvest shelf life of intact whole fruit and the cut fruit was, however, indicated by the similar decrease in the total volatiles as well as aliphatic and aromatic esters in stored NS and *Mission* fruit at 4 °C for a period of 24 h.

The loss of esters appears to be an important early reaction step related to loss of freshness during storage of cut cantaloupe melon. This reaction could potentially provide precursors for synthesis of other compounds that may further alter the typical cantaloupe melon flavor. The initial degradation of esters is unaffected by fruit storage temperature. When the cut fruit was kept at 22 °C, a similar rapid reduction in ester compounds occurred during the first day of storage (14). Refrigerated storage, however, favored the synthesis of terpenoid compounds in the fruit relative to storage at 22 °C.

The effects of storage and UV light on the volatile components of cut pineapple are similar to those that occurred in cut cantaloupe melon. The major compounds identified in pineapple using a similar GC-MS method are methyl 2-methylbutanoate, methyl hexanoate, methyl 5-hexanoate, ethyl hexanoate and ethyl 5-hexanoate (Table III; Figure 2). Some of the compounds, methyl and ethyl 3- methylthiopropionate, ethyl 2-methylbutanoate, and ethyl hexanoate are believed to be important contributors to the pineapple aroma *(23-25)*. Refrigerated storage of the fruit for a period of 24 h. caused a decrease of approximately 40% in total volatile and 50% in total esters relative to the freshly cut fruit. Exposure to ultraviolet light also had a similar degradative effect on the esters. In both cases, an increase in the amount of sesquiterpene phytoalexin, copaene occurred.

The production of esters is elaborated under strict biosynthetic control in the intact fruit by esterase (ES). Increase in ES activity occurs concurrently with the increased ester concentration during fruit maturation, reaching a peak towards the climacteric *(2)*. In cut fruits and vegetables, increased interaction of enzymes with their substrates occurs as a result of the loss of internal compartmentation that keeps them separate. The effect of storage at 4 °C on ES enzymes in fresh-cut cantaloupe melon and how this might affect loss of ester compounds in the fruit were determined. Enzymatic activity decreased during the first day of storage.(Table IV). Activity after 24 h of storage was 40% less than the freshly prepared fruit. ES activity between 24 and 72 h of storage increased 13% then remained relatively stable until 168 h of storage after which it increased.

Table III. Volatile compounds recovered in fresh-cut pineapple. Amounts are relative to the area of the internal standard. Peaks correspond to those identified in Figure 2.

Peak No	Compound	Relative Amount
1	ethyl acetate	0.1
2	ethyl propionate	0.03
3	methyl butanoate	0.19
4	methyl 2-methylbutanoate	1.02
5	ethyl 2- methylbutanoate	0.32
6	methyl hexanoate	2.27
7	methyl 5-hexenoate	3.25
8	ethyl hexanoate	1.24
9	ethyl 5-hexenoate	2.11
10	methyl 3(methylthio) propanoate	0.28
11	Z-ocimene	0.11
12	E-ocimene	0.62
13	3,7 -dimethyl-1,3,7 octatriene	0.53
14	ethyl 3 - (methylthio) propanoate	0.36
15	methyl (E) -4 octenoate	0.23
16	methyl octanoate	0.79
17	benzothiophene (standard)	1
18	methyl 5-acetoxyhexanoate	0.91
19	copaene	0.71

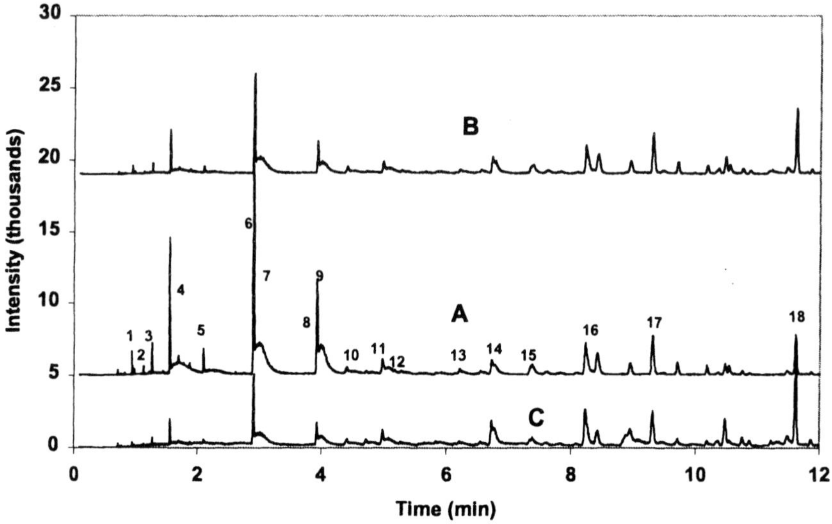

Figure 2. Total ion chromatogram of aroma compounds in fresh-cut pineapple (A), the cut fruit stored at 4 °C for 24 h (B) and fruit placed under UV radiation for 15 min.

Increase in ES activity should occur concurrently with loss of esters if ES initiated loss of esters is a rate limiting reaction step. The parallel loss of ES activity and esters with storage suggests that the ES involvement could be an early initiation of a reaction step, possibly at the time of processing, that then leads to the gradual loss of esters. Loss of esterase activity with storage time could be related to stability of the enzyme in the exposed tissue that is possibly catalyzed by protein degrading enzymes.

Table IV. Storate effect on fresh-cut cantaloupe melon esterase activity expressed as is the amount of protein in μg required to increase absorbance at 420 nm by 0.001.

Hours of Storage	Activity
0	836a
24	508b
72	603b
168	607b

Conclusions

Loss of esters occurs during storage of fresh-cut fruits. The reduction of ester contents in appears to be the part of the stress induced defense response to tissue disruption and exposure. Phytoalexin compounds with antimicrobial activities are produced concurrently with the loss of esters. The degradation of esters thus appears to be an important early reaction step related to loss of freshness during storage of fresh-cut fruits. Esterase activity is highest at the time of processing, which is when the initiation of ester hydrolysis could have occurred. ES activity decreases during refrigerated storage of cantaloupe melon. The rate of esterase activity during storage is unrelated to that of the loss of ester compounds.

References

1. Beaulieu J.C; Baldwin E.A. 2002. In *Fresh-cut Fruits and Vegetables: Science, Technology and Market.* Lamikanra, O. Ed. CRC Press, Boca Raton, FL **2002** pp 391-425.
2. Goodenough P.W; Riley T.E.W. 1985. *Ann. Appl. Biol.* **1985**, *107(2)*, 349-351.
3. Lamikanra, O.; Richard, O.A.; Parker, A. *Phytochemistry* **2002**, *60*, 27-32.
4. Kotseridis, Y.; Baumes, R.; Skouroumounis, G. K. J. Chromatog. A. **1998**, 824, 71-78.
5. Baldwin E.A.; Scott, J.W.; Einstein, M.A.; Malundo, T.M.M.; Carr, B.T.; Shewfelt, R.L.; Tandon, K.S. *J. Am. Soc. Hortic. Sci.* **1998**, *123*, 906-915.
6. Garcia, E.G. *Agronomia-Costarricense.* **1993**, *17*, 89-93.
7. Hildenbrand, S.; Ninnemann, H. *Physiol. Mol. Plant-Pathol.* **1994**, *44*, 335-347.
8. Megalla, S.E.; El-Keltawi, N.E.M.; Ross, S.A. *Herba Pol* **1980**, 26, 181-186.
9. Gamba-Invernizzi, A; Garlaschelli, L.; Rossi, A.; Vidari, G.; Vita-Finzi, P *J. Nat. Prod.* **1993**, *56,* 1948-1953
10. Chung, K.T..; Stevens, S.E. Jr.; Lin, W.F.; Wei, C.I. *Lett. Appl. Microbiol.* **1993**, *17*, 29-32.
11. Mueller, M.W.; Ishii H. *J. Phytopathol.* **1997**, *145*, 473-477.
12. Kagan-Zur V.; Tieman D.M.; Marlow, S.J.; Handa A.K.. *Plant Mol. Biol.* **1995**, *29* 1101-1110.
13. Kunzek, H.; Kabbert, R.; Gloyna, D. *Zeitschrift-fuer-Lebensmittel-Untersuchung-und -Forschung-A.* **1999**, *208*, 233-250

14. Lamikanra O, Richard O.A. *J. Agric. Food Chem.* **2002**, *50*, 4043-4047.
15. Bartley I.M.; Stoker,P.G.; Martin, A.D.E.; Hatfield, S.G.S; Knee, M. *J. Sci. Food Agric.* **1985**, *36*, 567-574.
16. Rowan D.D.; Lane, H.P.; Allen, J.M.; Fielder, S.; Hunt, M.B. *J. Agric. Food Chem.* **1996**, *44*, 3278-3285.
17. Anonimous. UC Vegetable Research and Information Center. Division of Agriculture and Natural Resources, University of California. **1999**, Publication 7218.
18. Cohen, R.A; Hicks, J.R. *J. Am Soc Hort Sci.*, **1986**, *111*, 553-557.
19. Ahmed E.A.; Eljack, A.E.; Mohamed Y.F. *Cucurbit Genetics Cooperative Report* **1997**, *20*, 30-31.
20. Pan R.S.; More T.A. *Euphytica* **1996**, *88*, 125-128.
21. Lamikanra O.; Juaraez, B.; Watson, M.A.; Richard, O.A.. *J. Sci. Food Agric.* **2003** (In press).
22. Nussbaumer, C.; Hostettler, B. In *Flavor Science: Recent Developments* A.J. Taylor and D.S. Mottram (Eds). Spec. Publ. R. Soc. Chem. Cambridge, UK. **1998**, 197, 70-73.
23. Takeoka, G.;, Buttery, R.G.; Flath, R.A.; Teranishi, R.; Wheeler, E.L.; Wieczorek, R.L.; Guentert, M. *ACS Symposium Series* R. Teranishi, R. Buttery and F. Shahidi (Eds.) **1991**, *388*, 223-237.
24. Buttery, R.G. In *Flavor Research: Recent Advances* S. Tannenbaum and P. Walstra (Eds.) Marcel Dekker, NY **1981**, 175-216.
25. Umano,K.; Hagi,Y.; Nakahara, K.; Shoji, A.; Shibamoto,T. *J. Agric. Food Chem.* **1992**, *40*, 599-603.

Indexes

Author Index

Subject Index

C

D

E

F

G

Q

S

T